INSIDER SECRETS TO HYDRAULICS

Insider Secrets to Hydraulics

Brendan Casey

First published in 2002 by
HydraulicSupermarket.com
PO Box 1029
West Perth WA 6872
Australia
Email: info@HydraulicSupermarket.com
Web: www.HydraulicSupermarket.com

Additional copies of this book can be purchased at:
www.InsiderSecretsToHydraulics.com

National Library of Australia Cataloguing-in-Publication entry:

Casey, Brendan.
Insider secrets to hydraulics.

Includes index.
ISBN 0 9581493 0 5.

1. Hydraulic machinery. 2. Hydraulic engineering - Equipment and supplies. I. Title.

621.2

Dedication

To my wife Jay, for her encouragement and support.

Contents

PART II "Houston we have a problem!" The Art & Science of Troubleshooting

PART III "Let the buyer beware" Insider Secrets to Repairing & Replacing Hydraulic Components

About the author

Brendan Casey is an author and fluid power consultant. He works with clients from a wide range of industries, to reduce the operating cost and increase the productivity of their hydraulic equipment. Brendan speaks, trains and consults on preventative maintenance, cost reduction, failure analysis, system design and procurement. To contact him, send email to bcasey@insidersecretstohydraulics.com or visit his company's Web site www.HydraulicSupermarket.com

Preface

Hydraulic equipment is used in hundreds of industries, from mining to aerospace, steel mills to food processing plants. According to industry sources, more than 70% of spare parts sold for hydraulic plant and machinery are used to replace defective components. The cause of 90% of these defects can be traced to improper operation or maintenance.

These statistics suggest that the fluid power industry profits from a lack of knowledge on the part of businesses that use hydraulic equipment. As an industry 'insider' for more than a decade, I can tell you that there are elements within the fluid power industry who deliberately exploit their customers' lack of knowledge to increase their profits.

I am not suggesting that the entire industry is crooked. However, there are unethical operators who will take advantage of you, if you give them an opportunity to do so. This book gives you the knowledge you need to avoid the rip-offs and reduce the operating cost of your hydraulic equipment.

WHO THIS BOOK IS FOR

I wrote this book for users of hydraulic equipment. It is intended to be an advocate for people who deal with the fluid power industry. Whether you're the owner/operator of a skid-steer loader or the maintenance manager for a large processing plant, you must read *Insider Secrets to Hydraulics*.

You don't need an engineering degree to read this book, although it will help if you have good mechanical aptitude and a basic understanding of hydraulics. A comprehensive Glossary has been included to make *Insider Secrets to Hydraulics* accessible to everyone with an interest in hydraulics.

WHAT YOU WILL LEARN

Insider Secrets to Hydraulics will show you how to slash the operating cost of your hydraulic equipment. You will learn:

- ➢ How to prevent costly, premature failure of hydraulic components.
- ➢ How to locate and rectify common hydraulic problems.
- ➢ How to avoid costly troubleshooting mistakes and rip-offs.
- ➢ How to save money on hydraulic component repairs.
- ➢ How to recognize and avoid common repair rip-offs.
- ➢ How to buy new hydraulic components at the lowest possible price.
- ➢ How to get free repairs after the warranty period has expired.

And there's much more! This book is filled with practical advice, supported by real examples that will save you time, money and aggravation.

WHAT YOU WILL NOT LEARN

Insider Secrets to Hydraulics is a unique insight into the fluid power industry. It is not intended to make you an expert in hydraulics and it does not cover the theory of fluid power hydraulics or explain in detail how different types of hydraulic components and systems work.

HOW TO USE THIS BOOK

You will get the greatest benefit from this book if you read it from beginning to end, rather than reading Chapters at random. Many of the Chapters build on the information contained in preceding Chapters, to construct a valuable body of knowledge.

UNITS OF MEASUREMENT

Metric units have been used in some examples for ease of explanation or calculation. Otherwise, US units are used throughout. Formulas for conversion between units can be downloaded from www.HydraulicSupermarket.com/technical

TELL ME WHAT YOU THINK

I would love to hear how you saved money using the tips and techniques contained in *Insider Secrets to Hydraulics.* And of course, if you have any suggestions for future editions, please send those too!

Email me at: author@insidersecretstohydraulics.com

Safety first

WARNING!

High-pressure fluid is present in operational hydraulic systems. Fluids under pressure are dangerous and can cause serious injury or death.

Always consider gravity before carrying out any work on a hydraulic machine. Any part of a hydraulic circuit that is supporting a load can remain pressurized after the system has been shut down. Attempting to remove a hydraulic line or component that is supporting a load can result in sudden release of pressure and/or uncontrolled movement of the load, causing serious personal injury and/or property damage.

Do not make modifications, repairs or adjustments to any hydraulic system unless you are competent or working under competent supervision. If in doubt, consult a qualified technician or engineer.

PART I

"Prevention is better than cure" The Science of Preventative Maintenance

1. Maintaining fluid cleanliness
2. Maintaining fluid temperature and viscosity within optimum limits
3. Maintaining hydraulic system settings to manufacturers' specifications
4. Scheduling component change-outs before they fail
5. Following correct commissioning procedures
6. Conducting failure analysis

INTRODUCTION

If you are serious about reducing the operating cost of your hydraulic equipment, you must start with an effective preventative maintenance program.

In the third Part of this book, I show you how to save money when repairing hydraulic components and purchasing new ones. But unless you have an effective preventative maintenance program in place, you will find yourself repairing hydraulic components and buying new ones more frequently than you should be.

Like most fluid power professionals, I have witnessed countless instances where expensive repairs and machine downtime could have been avoided, if the customer had implemented a simple and relatively inexpensive preventative maintenance program.

Six routines must be followed in order to minimize the chances of your hydraulic equipment suffering costly, premature component failures and unscheduled downtime:

- Maintain fluid cleanliness;
- Maintain fluid temperature and viscosity within optimum limits;
- Maintain hydraulic system settings to manufacturers' specifications;
- Schedule component change-outs before they fail;
- Follow correct commissioning procedures; and
- Conduct failure analysis.

Unfortunately, these things don't always take care of themselves. This means that some action is required to ensure that your hydraulic equipment operates under optimum conditions. In the following Chapters, I explain what you need to do to prevent premature failure of hydraulic components.

1

Maintaining fluid cleanliness

Maintaining fluid cleanliness involves defining a cleanliness level appropriate for the type of hydraulic system, sampling the fluid at regular intervals to monitor the actual cleanliness level against the target cleanliness level and taking remedial action as necessary, to achieve and maintain the target cleanliness level. Before I discuss this in detail, let me explain how contaminants in hydraulic fluid reduce the service life of hydraulic components.

FLUID CONTAMINATION AND ITS CONSEQUENCES

Contaminants of hydraulic fluid include insoluble particles, air, water or any other matter that impairs the function of the fluid. Air contamination can result in damage to hydraulic components through loss of lubrication, overheating and burning of seals. Water contamination can result in damage to hydraulic components through corrosion, cavitation and altered fluid viscosity.

Particle contamination accelerates wear of hydraulic components. The rate at which damage occurs is dependent on the internal clearances of the components within the system, the size and quantity of particles present in the fluid and system pressure. Typical internal clearances of hydraulic components are shown in exhibit 1.1.

Exhibit 1.1

COMPONENT TYPE	TYPICAL INTERNAL CLEARANCE IN MICRONS
Gear pump	0.5 - 5.0
Vane pump	0.5 - 10
Piston pump	0.5 - 5.0
Servo valve	1.0 - 4.0
Control valve	0.5 - 40
Linear actuator	50 - 250

Particles larger than a component's internal clearances are not necessarily dangerous. Particles the same size as the internal clearance cause damage through friction. But the most dangerous particles in the long-term are those that are smaller than the component's internal clearances. Particles smaller than 5 microns are highly abrasive. If present in sufficient quantities, these invisible 'silt' particles cause rapid wear, destroying hydraulic components.

To illustrate this point with an example, I was recently asked to conduct failure analysis on a piston pump with an expected service life of 10,000 hours. The unit in question had been removed from service after achieving only 2,000 hours. Examination revealed that this pump hadn't actually failed - it had been worn out through abrasion caused by contaminated fluid!

QUANTIFYING PARTICLE CONTAMINATION

Some level of particle contamination is always present in hydraulic fluid, even in new fluid. It is the size and quantity of these particles that we are concerned with. The level of contamination, or conversely the level of cleanliness, considered acceptable depends on the type of hydraulic system. Typical fluid cleanliness levels for different types of hydraulic systems, defined according to ISO, NAS and SAE standards, are shown in exhibit 1.2.

Exhibit 1.2

TYPE OF HYDRAULIC SYSTEM	MINIMUM RECOMMENDED CLEANLINESS LEVEL			MINIMUM RECOMMENDED FILTRATION LEVEL IN MICRONS ($\beta\chi \geq 75$)
	ISO 4406	NAS 1638	SAE 749	
Silt sensitive	13/10	4	1	2
Servo	14/11	5	2	3-5
High pressure (250-400 bar)	15/12	6	3	5-10
Normal pressure (150-250 bar)	16/13	7	4	10-12
Medium pressure (50 -150 bar)	18/15	9	6	12-15
Low pressure (< 50 bar)	19/16	10	-	15-25
Large clearance	21/18	12	-	25-40

ISO 4406 defines contamination levels using a somewhat complicated dual-scale numbering system. The first number refers to the quantity of particles larger than 5 microns per 100 milliliters of fluid and the second number refers to the number of particles larger than 15 microns per 100 milliliters of fluid.

The complicated part is that the quantities of particles these numbers represent are expressed as powers of the numeral 2. For example, a cleanliness level of 15/12 indicates that there are between 2^{14} (16,384) and 2^{15} (32,768) particles larger than 5 microns and between 2^{11} (2,048) and 2^{12} (4,096) particles larger than 15 microns, per 100 milliliters of fluid. A modified version of ISO 4406 includes 2 micron particle counts, in addition to the standard 5 micron and 15 micron counts.

New hydraulic fluid, straight from the drum, has a typical cleanliness level of 21/18. A 25 GPM pump operating continuously in fluid at this cleanliness level will circulate 3,500 pounds of dirt to the system's components each year! New fluid should always be filtered prior to use in a hydraulic system. The easiest way to do this is to pump the fluid into the reservoir through the return filter.

DEFINING A TARGET CLEANLINESS LEVEL

As an example, let's assume that we have a normal-pressure system and using exhibit 1.2 we define our target cleanliness level to be ISO 16/13. Having established the minimum fluid cleanliness level required for acceptable component life in this type of system, the next step is to monitor the actual cleanliness of the fluid to ensure that the target cleanliness level is maintained on a continuous basis. This involves taking fluid samples from the system at regular intervals and testing them for cleanliness.

TESTING FLUID CLEANLINESS

There are two ways of testing fluid cleanliness. The first involves sending a fluid sample to a laboratory for analysis. The lab results contain detailed information on the condition of the fluid. The information normally included in a fluid condition report, along with typical targets or alarm limits, are shown in exhibit 1.3.

Exhibit 1.3

CONDITION CATEGORY	RECOMMENDED TARGETS OR ALARM LIMITS
Fluid cleanliness level	Within targeted range chosen for the system or recommended by the manufacturer (ISO 4406)
Wear debris level	(Al) 5 ppm, (Cr) 9 ppm, (Cu) 12 ppm, (Fe) 26 ppm, (Si) 15 ppm
Viscosity	± 10 % of new fluid
Water content	< 100 ppm
Total Acid Number (TAN)	+ 25% of new fluid
Additive level	- 10% of new fluid

Oil analysis laboratories and some large equipment manufacturers, provide this service for a nominal, prepaid fee. The fee normally includes a sample container and postage to send the sample to the lab. Make sure that the lab you plan to use is able to specify the cleanliness level of the sample according to ISO 4406. It is also a good idea to state this requirement on the information card that accompanies the sample.

The second way to test a fluid's cleanliness level is to use a portable, electronic instrument designed for this purpose. This method is convenient and results are almost instant, however it shouldn't be considered a total substitute for lab analysis because the results do not include wear debris levels, viscosity, water content and other useful data. But when the two methods are used in combination, the frequency of lab analysis can be reduced. If you have a lot of hydraulic equipment to maintain, I recommend that you consider this option. For more information on fluid analysis and portable contamination-testing equipment, go to www.HydraulicSupermarket.com/fluidtest

Whichever method you use, it is important that the equipment you use to capture and contain the sample is absolutely clean. If you are taking multiple samples from different systems, take care not to cross-contaminate one fluid sample with another, and never take samples from drain plugs or other low-lying penetrations in the system, otherwise the results will be unreliable. Ideally, samples should be taken from the return line, upstream of the return filter, with the system at operating temperature.

There are no hard and fast rules on the service interval between samples, but assuming the results of the most recent sample were satisfactory, an interval of 500 hours is adequate for most systems. Consider shortening this interval for critical systems.

ACHIEVING A TARGET CLEANLINESS LEVEL

Going back to our example, let's assume that we have sampled the fluid in our system and received the fluid condition report. The report indicates an actual cleanliness level of ISO 19/16, well outside our target of 16/13. We know we are not going to get optimum service life from our system's components with this level of contamination in the fluid, so we need to fix it.

As you can see from exhibit 1.2, there is a correlation between fluid cleanliness level and the level of filtration in the system. Therefore, we need to check the system's current level of filtration. But first, let me explain filter ratings in more detail.

HYDRAULIC FILTER RATINGS

Hydraulic filters are rated according to the size of the particles they remove and the efficiency with which they remove them. Filter efficiency can be expressed either as a ratio (Beta, symbol β) for a given particle size (χ) or as a percentage. Filter Beta ratios and their corresponding efficiency percentages are shown in exhibit 1.4.

Exhibit 1.4

FILTER BETA RATIO AND PERCENTAGE EQUIVALENTS					
β	%	β	%	β	%
2.0	50.00	5.8	82.76	50.0	98.00
2.4	58.33	16.0	93.75	75.0	98.67
3.0	66.66	20.0	95.00	100.0	99.00
4.0	75.00	32.0	96.875	200.0	99.50

Filters are commonly classified according to *absolute* or *nominal* ratings. A filter that is classified *absolute* has an efficiency of 98% or better ($\beta\chi \geq 50.0$) at the specified micron size, and a filter that is classified *nominal* has an efficiency of between 50% and 95% ($\beta\chi 2.0$ - $\beta\chi 20.0$) at the specified micron size.

This can get a bit confusing, but the important thing to remember when purchasing filters for your hydraulic equipment, is that there is a significant difference in effectiveness between a 10-micron *nominal* and a 10-micron *absolute* filter element. Price and quality of filter elements can vary considerably. For more information on hydraulic filters and replacement elements, go to www.HydraulicSupermarket.com/filters

CHECKING THE FILTRATION LEVEL

According to exhibit 1.2, a filtration level of 10-micron with an efficiency of 98.67% ($\beta_{10} \geq 75$) is required to achieve a cleanliness level of ISO 16/13. This means that unless there is at least one filter in the system with a rating of 10-micron absolute, it is unlikely that a cleanliness level of 16/13 will be achieved, regardless of how many times the filters are changed. If a check of the existing filters reveals that this level of filtration is not present somewhere in the system, then either the level of filtration must be improved or the target cleanliness level must be revised downward.

I always install the highest practical level of filtration in the return line.The idea is that if the fluid in the reservoir starts clean and all fluid returning to the reservoir is adequately filtered, then the fluid cleanliness level will be maintained. This assumes of course, that the return filter element is changed before it clogs. If the element is allowed to clog, the bypass valve will open, allowing unfiltered fluid to enter the reservoir.

A word of warning here - don't automatically assume that the existing filter elements in a system can be automatically substituted with elements of a smaller micron size and/or higher efficiency.This will increase the restriction (pressure drop) across the filter and consequently the filter may no longer be able to handle its designed flow rate. If this happens, the filter's bypass valve will open and the filter will be ineffective. Filter manufacturers publish graphs that plot pressure drop against flow rate at a given fluid viscosity, according to an element's area, blocking size and efficiency.This information should be consulted before upgrading the elements in existing filter housings. For more information on filters and replacement elements, go to www.HydraulicSupermarket.com/filters

RECTIFYING ABNORMAL CONTAMINATION LOAD

Going back to our example, let's assume that the system's tank-top mounted return filter is rated 10-micron absolute ($\beta_{10} \geq 75$). Therefore, according to exhibit 1.2, our target cleanliness level of ISO 16/13 should be achievable with the existing level of filtration. So how do we explain the high level of particle contamination in the fluid?

If we are just starting our preventative maintenance program, this could be explained by a filter change that is long overdue. If we have some previous history on this system and the results of our last fluid sample were acceptable, we need to look for any abnormal source of contamination that is overloading the filters. Keep in mind that particle contamination can be generated internally or externally ingested.

Check the wear debris levels in the fluid condition report. This will indicate if the level of contamination being generated internally is abnormal. If wear debris levels are above alarm limits, this usually indicates that a component in the system has started to fail. Any metal-generating components need to be identified and changed-out.

Common entry points for externally ingested contamination are through the reservoir air space and on the surface of cylinder rods. Check that all penetrations into the reservoir air space are sealed and that the reservoir breather incorporates an air filter of 3-micron absolute or better. If the reservoir is not properly sealed and/or the breather not adequately filtered, dust can be drawn into the reservoir as the fluid volume changes.

Check that the chrome surfaces of all cylinder rods are free from pitting, dents and scores, and rod wiper seals are in good condition. Damaged cylinder rods and/or rod wiper seals allow dust that settles on the surface of the rod to enter the cylinder and contaminate the fluid.

Any increase in contamination load on the system's filters can result in premature clogging, which renders the filters ineffective. Early warning of filter clogging can be achieved by fitting filter-clogging indicators. These can be visual or electric, or a combination of both. The visual type has a button that pops up when the filter's bypass valve opens. The electric type has a set of contacts which close, usually operating a warning light on the machine's control panel when the bypass valve opens. Filter-clogging indicators aren't always fitted to filter housings as standard practice - but should be. For more information on filter-clogging indicators, go to www.HydraulicSupermarket.com/filters

FLUSHING THE FLUID

The next step is to change all of the filters in the system. Because our example system's current fluid cleanliness level of ISO 19/16 is well outside target, the fluid in the reservoir should be flushed before the filters are changed.This involves circulating the fluid in the reservoir through external filters for an extended period, or ideally, until the target cleanliness level is achieved.The equipment for doing this is commonly called a filter cart, which normally consists of an electric transfer pump and a set of filters mounted on a trolley.

The benefits of flushing the fluid in the system before changing the filters are that the system will be operating with cleaner fluid sooner, and the new filters don't have the job of cleaning up the fluid - they only have to maintain fluid cleanliness.

A filter cart can also be used to filter new fluid before it is added to a system. If you have a lot of hydraulic equipment to maintain, a filter cart is a wise investment. For more information on filter carts, go to www.HydraulicSupermarket.com/filtercarts

If you don't have access to a filter cart or it isn't practical to use one, purchase two sets of replacement filter elements at this time. Fit the first set immediately and replace them with the second set after 20 to 50 hours of service.The idea is that the first set of filters cleans the fluid and the second set keeps it clean. Either way, the fluid cleanliness level should be checked again after 50 hours of service to ensure the target cleanliness level has been achieved.

BENEFITS OF FLUID CONDITION MONITORING

The procedures for monitoring and maintaining fluid cleanliness described in this Chapter involve a continuous cycle of testing and corrective action. The benefits of regular fluid condition monitoring are illustrated in the following example.

Several years ago, I was responsible for a preventative maintenance program in a large, manufacturing plant. This plant operated 24 hours per day, 7 days a week. The manufacturing process was complex and highly integrated, such that a breakdown in one section of the plant would stop production across the whole plant. Consequently, unscheduled downtime was very costly in terms of lost production. As part of the preventative maintenance program, the fluid condition of the plant's 30 individual pieces of hydraulic equipment was closely monitored.

One day, as I was analyzing the latest batch of fluid condition reports, I noticed that one system was showing chromium levels way above normal. Investigation revealed that these high levels of chromium wear debris were being generated by a large diameter cylinder that had started to fail. The significance of the problem intensified when a check of the plant's spare parts inventory revealed that there was no spare on site and because the cylinder was unique to this piece of equipment, delivery time on a replacement was several weeks.

Early warning of this impending failure enabled a replacement cylinder to be manufactured and downtime to be scheduled for its change-out. This averted a long and costly period of unscheduled downtime. The management of this company needed no further convincing of the value of this aspect of the preventative maintenance program.

2

Maintaining fluid temperature and viscosity within optimum limits

Maintaining fluid temperature and viscosity within optimum limits involves defining an appropriate fluid operating temperature and viscosity range for the ambient temperature conditions in which the hydraulic system operates, and ensuring that both fluid temperature and viscosity are maintained within these defined limits. Before I discuss this in detail, let me explain the interrelationship of fluid temperature and viscosity, and how they impact upon hydraulic component life.

TEMPERATURE/VISCOSITY RELATIONSHIP OF HYDRAULIC FLUID

The viscosity of petroleum-based hydraulic fluid decreases as its temperature increases and conversely, viscosity increases as temperature decreases. This is why limits for fluid viscosity and fluid temperature must be considered simultaneously. Low fluid viscosity can result in component damage through inadequate lubrication caused by excessive thinning of the oil film, while excessively high fluid viscosity can result in damage to system components through cavitation.

Manufacturers of hydraulic components publish permissible and optimal viscosity values, which can vary according to the type and construction of the component. As a general rule, operating viscosity should be maintained in the range of 100 to 16 centistokes (460 to 80 SUS), however viscosities as high as 1000 centistokes (4600 SUS) are permissible for short periods at start up. Optimum operating efficiency is achieved with fluid viscosity in the range of 36 to 16 centistokes (170 to 80 SUS) and maximum bearing life is achieved with a minimum viscosity of 25 centistokes (120 SUS).

HYDRAULIC FLUID VISCOSITY GRADES

ISO viscosity grade (VG) numbers simplify the process of selecting a fluid with the correct viscosity for a system's operating temperature range. The VG numbers most commonly used in hydraulic systems are listed in exhibit 1.5.

Exhibit 1.5

ISO VISCOSITY GRADE NUMBER	TYPICAL AMBIENT OPERATING ENVIRONMENT
VG 22	Arctic climates
VG 32	Winter conditions in central Europe
VG 46	Summer conditions in central Europe
VG 68	Tropical climates / high ambient temperatures
VG 100	Extremely high ambient temperatures

A fluid's VG number represents its average viscosity in centistokes (cSt) at 40°C. For example, an ISO VG 32 fluid has an average viscosity of 32 centistokes at 40°C. Note that the average fluid viscosity of ASTM and BSI viscosity grade numbers are measured at 100°F (38.7°C). This means that fluids of a given ASTM or BSI grade are slightly more viscous than the corresponding ISO grade.

DETERMINING THE CORRECT VISCOSITY GRADE

In order to determine the correct fluid viscosity grade for a particular application, it is necessary to consider:

- starting viscosity at minimum ambient temperature;
- maximum expected operating temperature, which is influenced by maximum ambient temperature; and
- permissible and optimum viscosity range for the system's components.

In most cases, the machine manufacturer will specify the correct viscosity grade. It is important to understand that the machine manufacturer's recommended viscosity grade should change as the ambient temperature conditions in which the machine operates change.

I say this because several years ago I was involved in the analysis of several premature component failures from a mobile hydraulic machine. The machine was designed and built in the Northern Hemisphere, but was operating in high ambient air temperatures in the Southern Hemisphere. The components had failed due to inadequate lubrication, because of low fluid viscosity.

Investigation revealed that the fluid in the system was ISO VG 32. While this viscosity grade is suitable for cooler climates found in parts of the Northern Hemisphere, it was not suitable for the high ambient temperatures in which this machine was operating. The machine owner confirmed that the manufacturer's fluid recommendation was indeed ISO VG 32.

The machine manufacturer had not altered their fluid viscosity recommendation to take into account the higher ambient temperatures in which this particular machine was operating. This oversight resulted in several premature component failures because of low fluid viscosity.

The machine manufacturer's viscosity grade recommendation can be checked using the viscosity/temperature diagram shown in exhibit 1.6, assuming the minimum starting temperature and the hydraulic system's maximum operating temperature are known. For example, let's consider an application where the minimum ambient temperature is 15°C, the system's maximum operating temperature is 75°C, the optimum viscosity range for the system's components is between 36 and 16 centistokes and the permissible, intermittent viscosity range is between 1000 and 10 centistokes.

Exhibit 1.6

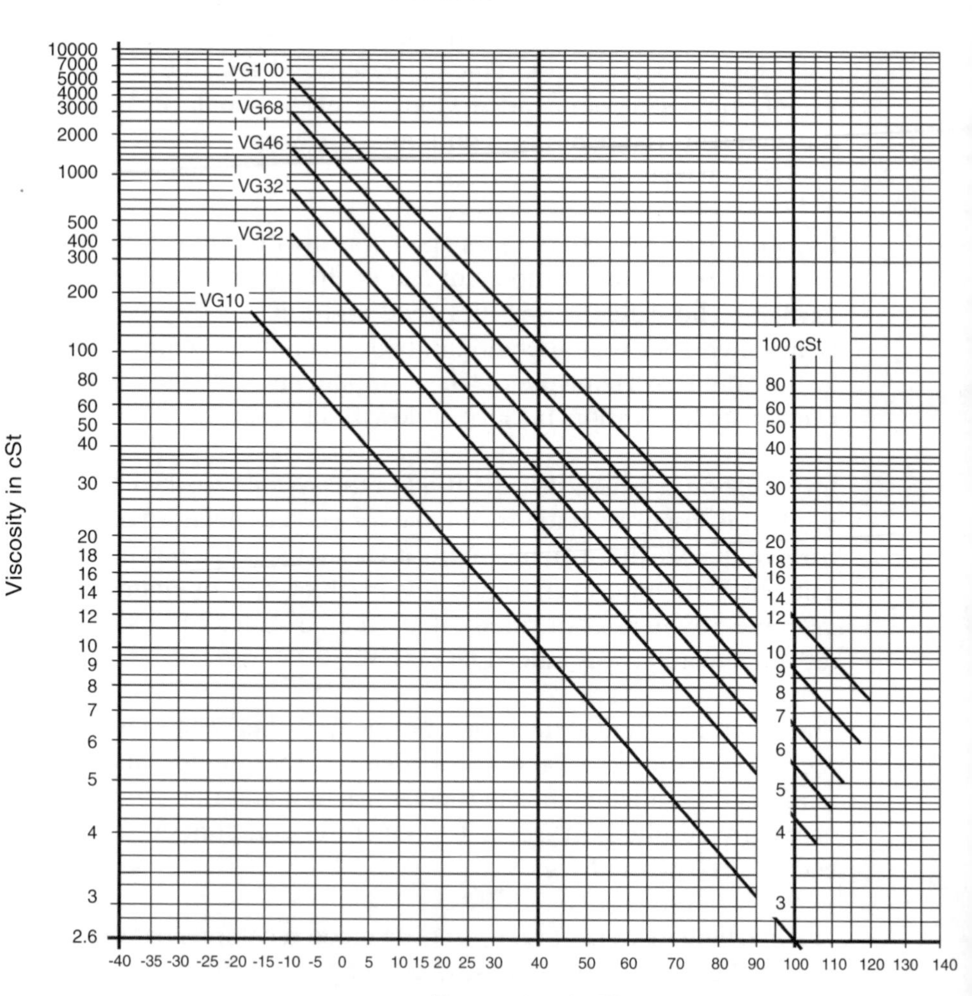

From the viscosity/temperature diagram in exhibit 1.6, it can be seen that to maintain viscosity above the minimum, optimum value of 16 centistokes at 75°C, an ISO VG 68 fluid is required. At a starting temperature of 15°C, the viscosity of VG 68 fluid is 300 centistokes, which is within the maximum permissible limit of 1000 centistokes at start up. If the machine manufacturer's recommendation was ISO VG 32 fluid under the same conditions, I would question it.

A word of warning here - do not change the fluid viscosity grade in a system without consulting the equipment manufacturer. Doing so may void the manufacturer's warranty and/or cause damage to the system's components.

DEFINING OPERATING TEMPERATURE LIMITS

Having established that the fluid in the system is the correct viscosity grade for the ambient temperature conditions in which the machine is operating, the next step is to define the fluid temperature equivalents of the optimum and permissible viscosity values for the system's components.

By referring back to the viscosity/temperature curve for VG 68 fluid in exhibit 1.6, it can be seen that an optimum viscosity range of between 36 and 16 centistokes will be achieved with a fluid temperature range of between 55°C and 78°C. The minimum viscosity for optimum bearing life of 25 centistokes will be achieved at a temperature of 65°C. The permissible, intermittent viscosity limits of 1000 and 10 centistokes equate to fluid temperatures of 2°C and 90°C, respectively.

Going back to our example, this means that with an ISO VG 68 fluid in the system, the optimum operating temperature is 65°C and maximum operating efficiency will be achieved by maintaining fluid temperature in the range of 55°C to 78°C. If cold start conditions at or below 2°C are expected, it will be necessary to pre-heat the fluid to avoid damage to system components. Intermittent fluid temperature in the hottest part of the system, which is usually the pump case, must not exceed 90°C.

Note that fluid temperatures above 82°C (180°F) damage seals, reduce the service life of the hydraulic fluid and in most cases, will cause the viscosity limits of the fluid to be exceeded. This means that the operation of any hydraulic system at temperatures above 82°C (180°F) is detrimental and should be avoided.

PREVENTING DAMAGE CAUSED BY HIGH TEMPERATURE OPERATION

To prevent damage caused by high fluid temperature and/or low fluid viscosity, a fluid temperature alarm should be installed in the system and all high temperature indications investigated and rectified immediately. The over-temperature alarm should be set to the temperature at which the minimum, optimum viscosity value is exceeded. As already explained, this will be dependent on the viscosity grade of the fluid in the system. In the example discussed above, the fluid temperature alarm would be set at 78°C.

Fluid temperature alarms are sometimes fitted by the manufacturer when the system is built, or they can be retrofitted. For more information on equipment for monitoring fluid temperature in hydraulic systems, go to www.HydraulicSupermarket.com/fluidtemp

RECTIFYING HIGH TEMPERATURE INDICATIONS

Going back to our example, let's assume that we have defined our system's viscosity/temperature limits and have fitted an over-temperature alarm to alert us when maximum fluid temperature and minimum fluid viscosity are exceeded. The system has an over-temperature indication and we know we are not going to get optimum component life with the system operating at this temperature, so we need to fix it.

High fluid temperature can be caused by internal or external factors, or a combination of the two. Abnormally high ambient temperature at the machine's location is an external factor that will reduce the hydraulic system's capacity to dissipate heat. Internal factors include anything that either reduces the system's capacity to dissipate heat or increases the heat load on the system.

The art of troubleshooting, which is covered in more detail in Part II, is a process of elimination that begins with checking the obvious things first. With this in mind, the first thing to check is the reservoir fluid level and, if low, fill to the correct level. Check that there are no obstructions to airflow around the reservoir, such as a build up of dirt or debris.

Inspect the heat exchanger and ensure that the core is not blocked. The ability of the heat exchanger to dissipate heat is dependent on the flow rate of both the hydraulic fluid and the cooling air or water circulating through the exchanger. Therefore, check the efficiency of cooling circuit components and replace as necessary.

When fluid moves from an area of high pressure to an area of low pressure without performing useful work, heat is generated. This means that any component in the hydraulic system that has abnormal, internal leakage will increase the heat load on the system and can cause the system to overheat. This could be anything from a cylinder that is leaking high-pressure fluid past its piston seal, to an incorrectly adjusted relief valve. Any heat-generating components need to be identified and changed-out.

Air generates heat when compressed. This means that air in the hydraulic fluid increases the heat load on the system and can cause the system to overheat. Air usually enters the system through the pump's inlet. Check the fluid level in the reservoir and, if low, fill to the correct level. Check pump intake lines are in good condition and all clamps and fittings are tight. In some systems, air can enter the pump through its shaft seal. Check the condition of the pump shaft seal and replace if necessary.

Cavitation occurs when the absolute pressure in any part of a hydraulic circuit falls below the vapor pressure of the hydraulic fluid. This results in the formation of vapor cavities within the fluid, which generate heat when compressed. Cavitation increases the heat load on the system and can cause the system to overheat. While cavitation can occur anywhere within a hydraulic circuit, it commonly occurs at the pump. If the pump has an inlet filter, check that it is not clogged. Check that the intake line between the reservoir and pump is not restricted.

Hydraulic systems that operate consistently above 85°C (185°F) usually have insufficient cooling capacity for the ambient temperature in which the machine is operating. In this case, additional cooling capacity will need to be installed. Consult the machine manufacturer or a fluid power engineer for guidance.

Continuing to operate a hydraulic system when the fluid is over-temperature is similar to operating an internal combustion engine with high coolant temperature. Damage is almost guaranteed. Therefore, whenever a hydraulic system starts to overheat, shut down the system, find the cause of the problem and fix it!

3

Maintaining hydraulic system settings to manufacturers' specifications

Maintaining hydraulic system settings to manufacturers' specifications involves regularly checking the operation and adjustment of the various circuit protection devices installed in a hydraulic system. This is in effect, a 'tune-up' of the hydraulic system. Before I discuss this in detail, let me explain how faulty or incorrectly adjusted circuit protection devices reduce hydraulic component life.

CONSEQUENCES OF A POORLY TUNED HYDRAULIC SYSTEM

Faulty or incorrectly adjusted circuit protection devices can result in reduced machine performance and cause damage to components through over-pressurization, cavitation and aeration. Over-pressurization occurs when the pressure developed in any part of a hydraulic circuit exceeds design limits. This condition is usually caused by faulty or incorrectly adjusted pressure control valves. Over-pressurization can result in burst hoses, blown seals and mechanical failure of pumps, motors, cylinders and valves.

Cavitation occurs when the volume of fluid demanded by any part of a hydraulic circuit exceeds the volume of fluid being supplied. This creates a partial vacuum within the circuit, which causes the fluid to vaporize. Cavitation causes metal erosion, which damages hydraulic components and contaminates the hydraulic fluid. In extreme cases, cavitation can result in mechanical failure of pumps and motors.

This condition can be caused by many factors, but the ones we are concerned with here are faulty or incorrectly adjusted devices that are designed to prevent cavitation, such as load or motion control valves and anti-cavitation valves.

Aeration occurs when air contaminates the hydraulic fluid. As explained in the previous Chapter, air can enter a hydraulic system through the pump inlet or shaft seal. Air can also enter a system through the gland of a double-acting cylinder because of faulty or incorrectly adjusted load control valves, float valves or anti-cavitation valves.

When a double-acting cylinder retracts under the weight of its load, the volume of fluid demanded by the rod side of the cylinder can exceed the volume of fluid supplied. As already explained, this condition normally results in cavitation. However, if a partial vacuum develops in the rod side of a double-acting cylinder, air can be drawn into the cylinder past its rod seals. This occurs because most rod seals are designed to keep high-pressure fluid in the cylinder and are not designed to keep air out. The result is aeration rather than cavitation. Both conditions cause damage to the cylinder. Aeration causes damage through loss of lubrication and overheating. When a mixture of air and oil is compressed in a cylinder, it can explode, damaging the cylinder and burning its seals. This is often referred to as the 'diesel effect' in reference to the combustion process in a diesel engine.

CHECKING AND ADJUSTING HYDRAULIC SYSTEM SETTINGS

Regular checking of hydraulic system settings not only ensures that the machine is operating efficiently, but also gives early warning of faulty circuit protection devices - before they cause component failures.

Machine manufacturers normally publish detailed instructions for checking and adjusting hydraulic system settings, which should be carried out with the hydraulic system at operating temperature. This information usually includes a list of equipment required to carry out the checks. A set of pressure gauges is often the only equipment required.

The installation of 'test-points' in relevant locations of the hydraulic circuit make checking pressures a simple task. A test-point is a device that enables a pressure gauge to be quickly connected to the circuit. The manufacturer may install the necessary test-points when the system is built, or they can be retrofitted.

More specialized test equipment, such as flow-testers, peak-pressure meters and multi-meters, is required to carry out checks and adjustments on complex systems or systems with electric or electronic controls. If you have a lot of hydraulic equipment to maintain, buying the necessary test equipment is a wise investment. For more information on test-points, pressure gauge sets, and other hydraulic test equipment, go to www.HydraulicSupermarket.com/hydtest

If you don't have access to the equipment required to check a hydraulic system's settings, you will need to hire a technician who has the equipment, to carry out the checks for you.

There are no hard and fast rules governing how often the settings of a hydraulic system should be checked, only guidelines based on common sense. Settings should always be checked at initial commissioning and when a hydraulic system is re-commissioned, following a component change-out or major maintenance work. I recommend checking settings every 1,000 to 2,000 hours, depending on how critical the system is and how vigilant you wish to be with your preventative maintenance program.

BENEFITS OF REGULARLY TUNING A HYDRAULIC SYSTEM

The benefits of regularly checking hydraulic system settings are best reinforced with an example. Several years ago, while working for a company that specialized in rebuilding hydraulic components, I was involved in the failure analysis of a large (and expensive) double-acting cylinder off a hydraulic excavator. This cylinder had been changed-out due to leaking rod seals after achieving only half of its expected service life. Analysis revealed that apart from the rod seals, which had failed as a result of the diesel effect, the other parts of the cylinder were in serviceable condition. As you now know, the diesel effect occurs in a cylinder when air is drawn past the rod seals, mixes with the hydraulic fluid and explodes when pressurized. Investigation revealed that the cause of the aeration and subsequent burning of the seals was a faulty float valve.

The function of a float valve on a hydraulic excavator is to allow the boom or arm to be lowered rapidly under its own weight. When activated, this valve connects the ports of the cylinder together, allowing the cylinder to retract under the weight of the boom or arm. The fluid displaced from the piston side of the cylinder is directed with priority to the rod side of the cylinder, before any excess volume is returned to the reservoir. An orifice controls the speed with which the cylinder retracts. If this valve malfunctions or is set incorrectly, a partial vacuum can develop in the rod side of the cylinder. Air enters the cylinder through the gland, resulting in 'dieseling' and subsequent failure of the rod seals.

If the operation of the float valve in the above example had been checked at regular intervals, the failure of this cylinder and the expense of its repair could have been avoided. This example also highlights the importance of checking the operation and setting of circuit protection devices whenever components are changed-out. As in this case, if the float valve is not identified as being faulty when a replacement cylinder is fitted, damage to this cylinder commences as soon as the machine is returned to service.

4

Scheduling component change-outs before they fail

Scheduling component change-outs before they fail involves determining the expected service life of each component in a system and scheduling their change-out when expected service life has been achieved, rather than running them to the point of failure. Before I discuss this in detail, let me explain the consequences of running hydraulic components until they fail and how this increases the operating cost of hydraulic equipment.

CONSEQUENCES OF COMPONENT FAILURES IN SERVICE

When a hydraulic component fails, large amounts of metallic particles are generated. These particles circulate in the hydraulic fluid, often causing damage to other components before the system's filters can remove them. In extreme cases, the contamination load can clog the filters, which results in unfiltered fluid being circulated through the system.

A component that fails in service is almost always more expensive to rebuild than a component that is removed from service in a pre-failed condition. A failure in service usually results in mechanical damage to the internal parts of the component. As a consequence, parts that would have been serviceable have to be replaced. In extreme cases, components that would have been economical to repair become uneconomical to repair, increasing the cost of component replacement by as much as 40%.

DETERMINING EXPECTED SERVICE LIFE

The expected service life of individual components within a hydraulic system varies and is influenced by a number of factors. These include the type and construction of the component, circuit design, operating load and duty-cycle. Machine manufacturers determine the expected service life of components within a particular system, by considering these variables in combination with historical data on achieved service life. This information is normally available from machine manufacturers upon request. To minimize the chances of hydraulic components failing in service, the machine manufacturers' recommendations on expected service life should be used to schedule component change-outs.

BENEFITS OF SCHEDULING COMPONENT CHANGE-OUTS

The benefits of scheduling component change-outs upon completion of their expected service life are best reinforced with an example. A manufacturing company recently hired me to check the performance of four piston pumps operating a large press. The pumps had clocked over 10,000 hours in service and the customer's concern was that if pump performance was down, production would be too.

The test results revealed that the performance of all four pumps was within acceptable limits. In my report, I advised the customer that there would only be a minimal increase in productivity if the pumps were replaced. I further advised that the change-out of all four pumps should be scheduled immediately!

The foundation for this recommendation was that the pumps had exceeded their expected service life and because of this, there was a significant risk of a bearing failure. I explained to the customer that if a bearing failed in any of the pumps, not only would the cost to rebuild the pump increase, other components in the system could be damaged.

The customer took my advice, but unfortunately, a bearing failed in one of the pumps before all of the change-outs were completed. A piece of cage from the failed bearing found its way into the main press cylinder, causing $5,000 damage. The pump that failed cost 50% more to rebuild than the three units that were removed from service in pre-failed condition.

The additional repair costs in this case were significant and could have been avoided, if the pumps had been changed-out as soon as they achieved their expected service life. In most cases, running hydraulic components beyond their expected service life is false economy.

5

Following correct commissioning procedures

Following correct commissioning procedures ensures that hydraulic components are commissioned properly during installation or when a system is re-started after maintenance. Improper commissioning is one of the most common causes of premature failure of hydraulic components.

CONSEQUENCES OF IMPROPER COMMISSIONING

Incorrect commissioning during start-up can result in damage through inadequate lubrication, cavitation and aeration. In many cases, this damage will not show itself until the component fails, hundreds or even thousands of service hours after the event.

A common misconception among maintenance personnel with limited training in hydraulics, is that because oil circulates through hydraulic components in operation, no special attention is required during installation beyond fitting the component and connecting its hoses. Nothing could be further from the truth, as the following example illustrates.

I was asked recently to conduct failure analysis on a piston motor that was the subject of a warranty claim. The motor had failed after only 500 hours in service, 7,000 hours short of its expected service life. Failure analysis was conclusive. The motor's bearings had failed through inadequate lubrication, as a result of the motor being started with insufficient fluid in its case.

The motor's case should have been filled with clean hydraulic fluid prior to start-up. Starting a piston pump or motor without doing so, is similar to starting an internal combustion engine with no oil in the crankcase - premature failure is almost guaranteed.

It is standard practice within the fluid power industry for improper commissioning to void warranty and in this particular example, the warranty claim was rejected on this basis. Consequently, the customer was lumbered with an expensive repair bill that could have been easily avoided.

HYDRAULIC SYSTEM COMMISSIONING PROCEDURES

To prevent damage to hydraulic components during initial start-up and subsequent exclusion of warranty, obtain and follow the machine manufacturers' commissioning procedures. If the manufacturers' machine-specific commissioning procedures are not available, the following general procedures should be used:

Pre-start

If the system is down as a result of a major component failure:

- Drain and clean the reservoir, to ensure that it is free from metallic debris and other contamination. Failure to do so may result in damage to the pump(s) and/or other components on start-up.
- Change all the filters.
- Change the fluid. On large systems, where the cost of changing the fluid may be prohibitive, the fluid should be flushed until a cleanliness level of ISO 4406 16/13 or better is achieved.

When installing pumps and motors, check the drive coupling for fit on the pump or motor shaft. Loose fitting couplings cause accelerated wear of the drive shaft and should be replaced.

On closed-circuit systems (hydrostatic transmissions), closely inspect the high-pressure hoses or pipes between the pump and motor, and replace any suspect lines. A burst hose or pipe in service can result in the destruction of the pump and/or motor through cavitation.

Cylinders - Fill cylinders with clean hydraulic fluid through their service ports, before connecting service lines. This reduces the risk of air compression within the cylinder (dieseling) on start-up, which will result in damage to the cylinder and its seals.

Motors - Fill the case of piston-type motors with clean hydraulic fluid through the highest case drain port and connect the case drain line. Failure to do so will result in damage to the motor through inadequate lubrication on start-up. Units that are mounted vertically, with the shaft up, require special attention to ensure that the fluid level in the case is high enough to lubricate the front shaft bearing(s).

Pumps - After installing the pump(s) and connecting service lines:

- Open the intake line isolation valve at the reservoir.
- On pumps with a flooded inlet, i.e. pump inlet is below reservoir fluid level, carefully loosen the intake line fitting at the pump, to allow trapped air to escape. This ensures that the intake line is full of fluid. This step is not necessary with piston-type pumps that have a flooded housing (see below).
- On piston-type pumps fitted with an external case drain line, fill the pump case with clean hydraulic fluid through the highest case drain port and connect the case drain line. Failure to do so will result in damage to the pump through inadequate lubrication on start-up. Units that are mounted vertically with the shaft up require special attention, to ensure

that the fluid level in the case is high enough to lubricate the front shaft bearing(s).

- On piston-type pumps with a flooded housing, i.e. pump case and inlet are connected internally and are below reservoir fluid level, carefully loosen the uppermost plug in the pump case to allow trapped air to escape. This ensures that the case is full of fluid. Failure to do so will result in damage to the pump through inadequate lubrication on start-up.
- On closed-circuit pumps (hydrostatic transmissions) install a 0-900 PSI pressure gauge in the charge circuit - refer to the machine manufacturer's instructions for guidance.

Start-up

- Check all pipe and hose connections are tight.
- Confirm reservoir fluid level is above minimum.
- **CAUTION! Confirm all controls are in neutral to ensure that the system will start unloaded. Take safety precautions to prevent machine movement, in the event that the system is activated during initial start-up.**
- If the prime mover is electric, momentarily start and then stop the electric motor to visually confirm that the direction of motor rotation is correct for the pump. Rotating the pump in the wrong direction can damage the pump.
- Start the prime mover and run at the lowest possible rpm.
- On closed-circuit systems (hydrostatic transmissions) monitor the pressure gauge previously installed in the charge circuit. If the manufacturer's specified charge pressure, typically 110 to 360 PSI, is not established within 20 to 30 seconds, shut down the prime mover and investigate the problem. Do not operate the system without adequate charge pressure - damage to the transmission pump and/or motor will result.
- On variable-displacement pumps and motors with external, low-pressure pilot lines, carefully loosen the pilot line fitting at the pump or motor to allow trapped air to escape. This ensures that the pilot line is full of fluid. **CAUTION! Do not**

bleed pilot lines carrying high-pressure fluid. Personal injury may result. If in doubt, do not bleed pilot lines!

- Allow the system to run at idle and unloaded for ten minutes. Monitor pump(s) for unusual noise or vibration, inspect system for leaks and observe reservoir fluid level.
- Function the system without load. Stroke cylinders slowly, taking care not to develop pressure at the end of stroke, to avoid compression of trapped air, which will result in damage to the cylinder and its seals through the diesel effect. Continue to function the system until all air is expelled and all actuators operate smoothly.
- With the system at operating temperature, check and if necessary adjust settings of circuit protection devices according to manufacturers' specifications.
- Function-test the system under load.
- Inspect the system for leaks.
- Shut down prime mover, remove all gauges fitted during commissioning, check reservoir fluid level and, if necessary, fill to the correct level.
- Return machine to service.

6

Conducting failure analysis

You're probably wondering what failure analysis has to do with preventative maintenance. You might even be thinking that the two concepts are contradictory. In fact, failure analysis is an important preventative measure and an essential element of any preventative maintenance program. The logic for this is simple - if a component fails prematurely and the cause of the failure is not identified and rectified immediately, then the replacement component is likely to suffer a similar fate.

The objective of a preventative maintenance program is to reduce the chances of your hydraulic equipment suffering costly, premature component failures and unscheduled downtime. Even with the best preventative maintenance program, a premature failure can still occur. Components can also fail prematurely for reasons that are unrelated to the preventative measures discussed in the preceding Chapters. Manufacturing defects, circuit design faults and operator abuse are typical examples.

When a premature failure does occur, it is essential that a thorough analysis be conducted in order to determine the cause.

Conducting failure analysis on hydraulic components is a highly specialized task that requires a detailed understanding of hydraulic circuits, the construction of hydraulic components and their modes of failure.

Reputable hydraulic repair shops provide this service for a nominal fee, which in most cases will be included in their price to rebuild the component. If a new component has been fitted and the failed component is not going to be repaired, failure analysis should still be carried out. In this case, you could ask the supplier of the new component to analyze the failed component for you. Be aware however, that the quality of any failure analysis is dependent on the expertise of the technician or engineer conducting the analysis. Not all hydraulic shops have this expertise in-house. If in doubt, consult a specialist in this area.

Failure analysis may not be conclusive in all cases, but it can provide valuable clues to identifying the cause of failure. Establishing the cause of failure is essential, so that remedial action can be taken to prevent similar failures.The benefits of conducting failure analysis should be evident from any of the examples described in the preceding Chapters. It will save you time, money and a lot of aggravation!

PART I CONCLUSION

The preventative maintenance program outlined in the preceding Chapters requires time, effort and some expense to implement. But it is very cost-effective. This investment is quickly recovered through savings as a result of improved machine performance, increased component life, increased fluid life, reduced downtime and fewer repairs.

PART II

"Houston we have a problem!" The Art & Science of Troubleshooting

INTRODUCTION

Troubleshooting hydraulic systems can be a complex exercise. It involves a lot of science and sometimes, a bit of art. As I said in the Preface, this book is not intended to make you an expert in hydraulics, but it will show you how to save a lot of money on the operation and maintenance of your hydraulic equipment.

With this in mind, the objective of the following Chapters is to explain the fundamentals of troubleshooting, so that when you do have a problem with your hydraulic equipment, you can be sure of two things. Firstly, that you have carried out an informed assessment of the problem and eliminated all of the obvious causes before you incur the expense of hiring a technician. And secondly, that if you do need to hire a technician, you are able to evaluate the technician and his diagnosis so you don't end up paying for his on-the-job training, or fall victim to a troubleshooting rip-off.

7

Troubleshooting basics

Troubleshooting is a process of elimination that begins with checking the obvious things first. In order for the 'obvious things' to be obvious, an understanding of the fundamental laws of hydraulics is required. The following must be kept in mind when troubleshooting a hydraulic system:

- Hydraulic pumps create flow - not pressure.
- Resistance to flow creates pressure.
- Flow determines actuator speed.
- Pressure determines actuator force.
- Fluid under pressure takes the path of least resistance.
- When fluid moves from an area of high pressure to an area of low pressure (pressure drop) without performing useful work, heat is generated.

For those of you who wish to delve deeper into the theory of fluid power hydraulics, a good textbook is a sound investment. There are many excellent books on the subject and a list of my favorites can be found at www.HydraulicSupermarket.com/books

TROUBLESHOOTING 101 – THEORY AND PRACTICE

Theory is great, but it always makes more sense when put into practice. Let me explain how the fundamental laws of hydraulics are applied in a troubleshooting situation. A few years ago, I was called to a problem on a mobile hydraulic machine. The pump had failed and been replaced with a rebuilt unit. The message I received from the customer was "...the replacement pump is not developing pressure."

If the customer had read this book, he would have known that this statement is misleading. We know that a pump can only produce flow and, assuming the pump in question was producing flow, an absence of pressure indicates an absence of resistance to flow.

After a quick, visual check of the reservoir fluid level and the pump's installation (never overlook the obvious) I was able to establish that the pump was producing flow. Having established this fact and knowing that fluid under pressure always takes the path of least resistance, I was now looking for the point at which pump flow was escaping from the circuit.

Knowing that heat is generated when there is a pressure drop, I operated the system while checking the temperature of individual components using an infrared heat gun. Within a few minutes, the relief valve had become the hottest part of the system. This indicated that the relief valve was the problem.

I increased the pressure setting of the relief valve, but no increase in system pressure resulted. It is important to note that a logical interpretation of this could have been that the relief valve was not the problem. If flow <u>is</u> passing over a relief valve and its pressure setting is increased, system pressure should increase as a result of an increase in resistance to flow. This logic is based on the assumption that the relief valve is serviceable.

Because I had already proved that the relief valve was generating heat, using an infrared heat gun, my interpretation was that the relief valve was faulty. I shut down the system and disassembled the relief valve. It turned out that a piece of debris from the failed pump had found its way into the relief valve and was holding the valve off its seat! (Remember what I said in Chapter 4 about scheduling component change-outs before they fail?)

The type and variation of problems a hydraulic system can encounter are infinite and therefore it is not possible to cover them all here. But as you can see from this example, a solid understanding of the fundamental laws of hydraulics can be applied in any situation, and is the key to effective troubleshooting.

Manufacturers' technical literature, especially machine-specific troubleshooting information, can also be of assistance and should be referred to where available.

8

Symptoms of common hydraulic problems and their causes

When a hydraulic system has a problem, common symptoms are abnormal noise, high fluid temperature and slow operation. These symptoms can appear separately or in combination. Before I discuss these in detail, let me explain what I mean when I talk about a symptom, a problem and an underlying cause.

A symptom is something that alerts you to the fact that something is wrong. A problem is something that causes the symptom, and an underlying cause is the root of the problem. To give you an example, consider a situation where a hydraulic system is abnormally noisy, as a result of cavitation caused by a collapsed pump intake line. Abnormal noise is the symptom, it alerts you that there is something wrong. Cavitation is the problem, it is causing the abnormal noise. The collapsed pump intake line is the underlying cause, it is the root of the problem.

ABNORMAL NOISE

Abnormal noise in hydraulic systems is usually caused by aeration or cavitation. Aeration occurs when air contaminates the hydraulic fluid. Air in the hydraulic fluid makes an alarming banging or knocking noise when it compresses and decompresses, as it circulates through the system. If aeration is the problem, the fluid in the reservoir will have a frothy appearance and actuator movement will usually be erratic.

Air usually enters the hydraulic system through the pump's inlet. Check pump intake lines are in good condition and all clamps and fittings are tight. Flexible intake lines sometimes become porous with age, therefore replace old or suspect intake lines. If the fluid level in the reservoir is low, a vortex can develop, allowing air to enter the intake line. Check the fluid level in the reservoir and, if low, fill to the correct level. In some systems, air can enter the pump through its shaft seal. Check the condition of the pump shaft seal and if it is leaking, replace it.

Aeration reduces the service life of the hydraulic fluid and causes damage to system components through loss of lubrication, overheating and burning of seals. To prevent aeration caused by a vortex at the pump intake, fit a low fluid level alarm to the reservoir. This is particularly important in the case of mobile hydraulic machines, where the reservoir fluid level can be affected by the incline of the machine. Low fluid level alarms are sometimes fitted by the manufacturer when the system is built, or they can be retrofitted. For more information on equipment for monitoring reservoir fluid level, go to www.HydraulicSupermarket.com/fluidlevel

Cavitation occurs when the volume of fluid demanded by any part of a hydraulic circuit exceeds the volume of fluid being supplied. This causes the absolute pressure in that part of the circuit to fall below the vapor pressure of the hydraulic fluid. This results in the formation of vapor cavities within the fluid, which implode when compressed, causing a characteristic knocking noise.

While cavitation can occur just about anywhere within a hydraulic circuit, it commonly occurs at the pump. A clogged inlet filter or restricted intake line will cause the fluid in the intake line to vaporize. If the pump has an inlet filter, check that it is not clogged. If a gate-type isolation valve is fitted to the intake line, check that it is fully open. This type of isolation device is prone to vibrating closed. Check that the intake line between the reservoir and pump is not restricted. Flexible intake lines are prone to collapsing with age, therefore replace old or suspect intake lines.

The consequences of cavitation in a hydraulic system can be very serious. Cavitation causes metal erosion, which damages hydraulic components and contaminates the hydraulic fluid. In extreme cases, cavitation can result in major mechanical failure of pumps and motors. To prevent damage caused by cavitation, shut down the system immediately when any abnormal noise is detected, and do not operate the machine until the cause of the problem has been identified and fixed.

HIGH FLUID TEMPERATURE

Fluid temperatures above 82°C (180°F) damage seals and reduce the service life of hydraulic fluid. This means that the operation of any hydraulic system at temperatures above 82°C is detrimental and should be avoided. Fluid temperature is too high when viscosity falls below the optimum value for the system's components. As explained in Chapter 2, the temperature at which this problem occurs is dependent on the viscosity grade of the fluid in the system and can be well below 82°C.

High fluid temperature can be caused by anything that either reduces the system's capacity to dissipate heat or increases the heat load on the system. Hydraulic systems dissipate heat through the reservoir. Therefore, check the reservoir fluid level and, if low, fill to the correct level. Check that there are no obstructions to airflow around the reservoir, such as a build up of dirt or debris.

Inspect the heat exchanger and ensure that the core is not blocked. The ability of the heat exchanger to dissipate heat is dependent on the flow rate of both the hydraulic fluid and the cooling air or water circulating through the exchanger. Therefore, check the performance of cooling circuit components and replace as necessary.

The performance of the cooling fan or water pump can be checked by measuring the cooling fan or water pump speed and comparing actual rpm against design rpm. This is easily checked using a photocell tachometer. A photocell tachometer uses a light beam aimed at a reflective mark placed on a rotating device to measure rpm. No contact with the rotating shaft is required. For more information on photocell tachometers, go to www.HydraulicSupermarket.com/tachometers

When there is a pressure drop, heat is generated. This means that any component in the system that has abnormal, internal leakage will increase the heat load on the system and can cause the system to overheat. This could be anything from a cylinder that is leaking high-pressure fluid past its piston seal, to an incorrectly adjusted relief valve. Identify and change-out any heat-generating components.

A word of warning here - the temperature of hydraulic components can exceed 90°C (194°F), therefore do not try to identify heat-generating components by touch. Serious burns can result. Use a temperature-measuring instrument such as an infrared heat gun, also called an infrared thermometer, to identify components that are generating heat. An infrared thermometer is a hand-held device that measures temperature using non-contact infrared technology. For more information on temperature measuring instruments, go to www.HydraulicSupermarket.com/thermometers

A common cause of heat generation in closed center circuits is the setting of relief valves below, or too close to, the pressure setting of the variable-displacement pump's pressure compensator. This prevents system pressure from reaching the setting of the pressure compensator. Instead of pump displacement reducing to zero, the pump continues to produce flow, which passes over the relief valve, generating heat. To prevent this problem in closed center circuits, the pressure setting of the relief valve(s) should be 250 PSI above the pressure setting of the pump's pressure compensator - see exhibit 2.1.

Exhibit 2.1

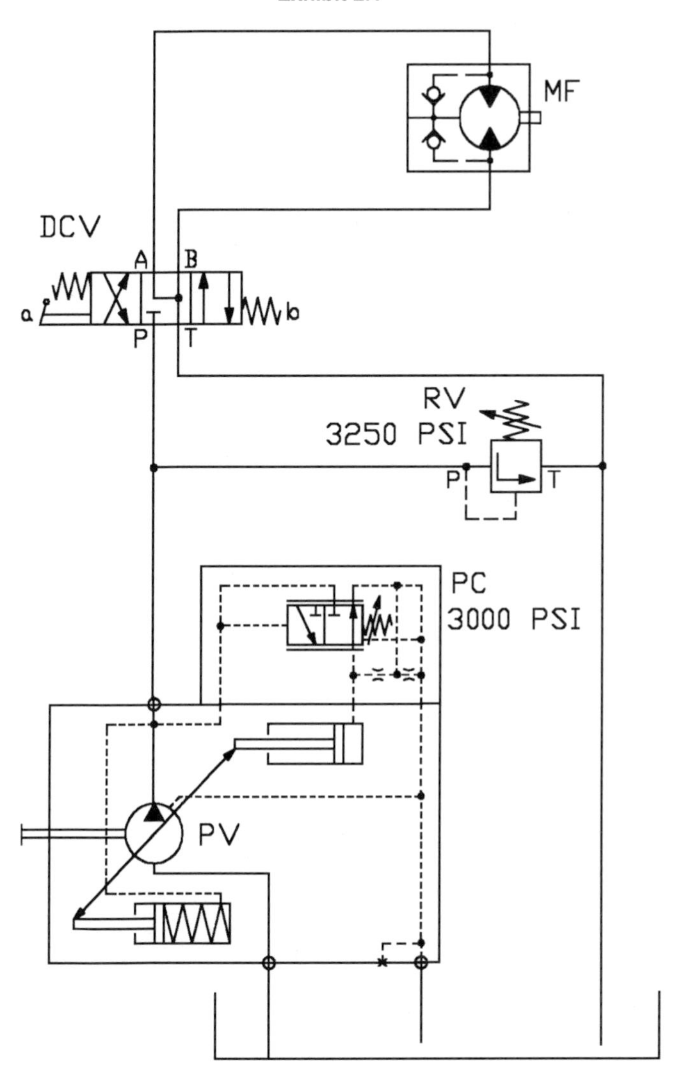

Closed center circuit showing relief valve (RV) setting 250 PSI above the pressure compensator (PC) setting of the variable pump (PV).

Air generates heat when compressed. This means that aeration increases the heat load on the hydraulic system and can cause the system to overheat. Therefore, inspect the system for possible causes of aeration.

As already explained, cavitation results in the formation of vapor cavities in the hydraulic fluid. These cavities generate heat when compressed. Cavitation increases the heat load on the hydraulic system and can cause the system to overheat. Therefore, inspect the system for possible causes of cavitation.

Hydraulic systems that operate consistently above 85°C (185°F) usually have insufficient cooling capacity for the ambient temperature in which the machine is operating. In this case, additional cooling capacity will need to be installed. Consult the machine manufacturer or a fluid power engineer for guidance.

In addition to damaging seals and reducing the service life of the hydraulic fluid, high fluid temperature can cause damage to system components through inadequate lubrication as a result of excessive thinning of the oil film (low viscosity). To prevent damage caused by high fluid temperature, a fluid temperature alarm should be installed in the system and all high temperature indications investigated and rectified immediately. Fluid temperature alarms are sometimes fitted by the manufacturer when the system is built, or they can be retrofitted. For more information on equipment for monitoring fluid temperature in hydraulic systems, go to www.HydraulicSupermarket.com/fluidtemp

SLOW OPERATION

A reduction in machine performance is often the first indication that there is something wrong with the hydraulic system. This usually manifests itself in longer cycle times or slow operation. Remember that in a hydraulic system, flow determines actuator speed. Therefore, a loss of speed indicates a loss of flow. The objective of the troubleshooting task then, is to locate where flow is being lost from the system.

Flow can escape from the hydraulic circuit through external or internal leakage. External leakage such as a burst hose is usually obvious and therefore easy to find. Internal leakage can occur in the pump, valves or actuators and, unless you are gifted with x-ray vision, is more difficult to isolate. Locating internal leakage is the subject of the next Chapter.

Where there is internal leakage there is a pressure drop, and where there is a pressure drop heat is generated. This increases the heat load on the hydraulic system and therefore slow operation and high fluid temperature often appear together. This can be a vicious circle. When fluid temperature increases, viscosity decreases. When viscosity decreases, internal leakage increases. When internal leakage increases, heat load increases, resulting in a further increase in fluid temperature and so the cycle continues.

9

Locating internal leakage

Where there is internal leakage there is a pressure drop and where there is a pressure drop, heat is generated. This makes an infrared heat gun a useful tool for identifying components that are leaking internally. But in some cases, temperature measurement will not be conclusive in isolating the source of the leakage and it will be necessary to use a flow-tester for this purpose.

A flow-tester is a portable instrument that comprises a flow turbine for measuring flow rate, an adjustable orifice that is used to increase the resistance to flow (load valve) and a pressure gauge, which measures pressure upstream of the load valve. When connected into the circuit, the flow-tester allows flow rate to be monitored while the resistance to flow (and therefore pressure) is increased using the load valve.

LOCATING INTERNAL LEAKAGE USING A FLOW-TESTER

Using a flow-tester in a troubleshooting situation involves a process of elimination. The point at which the flow-tester is connected into the circuit determines the conclusions that are drawn. To illustrate this point, consider a very simple hydraulic circuit comprising only four components of interest: a fixed-displacement pump (PF); a relief valve (RV); a directional control valve (DCV) and a double-acting cylinder - see exhibit 2.2. The pump has a rated flow of 10 GPM and the pressure setting of the relief valve is 3000 PSI.

Exhibit 2.2

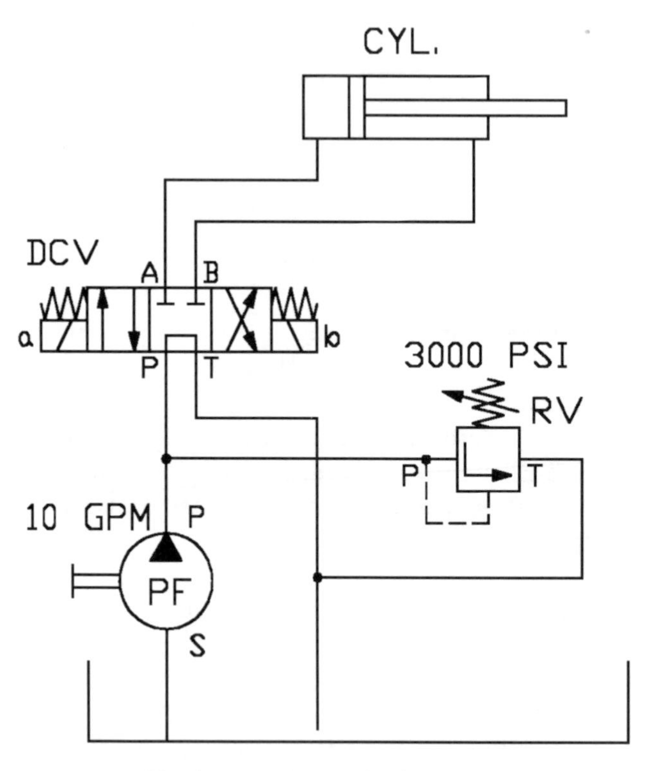

Simple open center circuit showing
four components of interest.

We have noticed that cylinder speed is slowing under load. We know that flow determines speed and fluid under pressure takes the path of least resistance. Therefore, we can assert that when the load comes on the cylinder, some of the available flow is taking an easier path back to the reservoir. The question is, which of the four components in the system is allowing this to happen?

Let's say we have a hunch that fluid is leaking past the piston seal of the cylinder. In this case, the flow-tester would be connected into the circuit after the pump, relief valve and directional control valve but before the cylinder. To do this, we remove the two hoses from the service ports of the cylinder and connect these hoses directly to the flow-tester. This takes the cylinder out of the circuit for the purposes of this test.

Because we have isolated the cylinder, there are only two possible conclusions that we can draw from this particular test. Either the cylinder is leaking internally or the cylinder is not leaking internally. As you can see, this is an exercise in logic.

To conduct this test, the directional control valve must be activated to direct flow to the flow-tester. **A word of warning here** - some flow-testers are only designed to work in one direction i.e. they have a specific inlet and outlet port (unidirectional). Assuming our hypothetical flow-tester is unidirectional, we need to ensure that when the directional control valve is activated, it is directing flow to the inlet port of the flow-tester.

With the prime mover running and the directional control valve activated, pump flow circulates through the flow-tester and back to the reservoir. Because the load valve on the flow-tester is fully open, there is little resistance to flow and therefore minimal pressure is developed. At this point, we note the flow and pressure readings on the flow-tester, which for the purposes of this example are 9.8 GPM and 50 PSI. It is a good idea to record these readings as the test is conducted. To proceed with the test we increase the resistance to flow (and therefore pressure) using the load valve on the flow-tester, while taking note of the flow rate.

A word of warning here - when a flow-tester is used to load a hydraulic system, there is a pressure drop across the load valve. As you now know, when there is a pressure drop heat is generated. This means that when loaded, the flow-tester will heat the hydraulic fluid and may cause the hydraulic system to overheat if it is left loaded for a prolonged period. For this reason, the duration of any flow test should be kept as short as possible.

Let's say the results of this test look like those shown in exhibit 2.3. As you can see, the flow available to the cylinder is 9.5 GPM at 500 PSI, decreasing to 5.9 GPM at 2500 PSI. This represents a 38% reduction in flow between 500 PSI and 2500 PSI (9.5 - 5.9 = 3.6 & 3.6 ÷ 9.5 x 100 = 38). This means that we would expect to see a 38% reduction in cylinder speed between 500 and 2500 PSI! This tells us that the cylinder is not the cause of the problem. Unfortunately, what these results do not tell us is which one of the other three components is causing the problem.

Exhibit 2.3

PRESSURE PSI	FLOW RATE GPM
50	9.8
500	9.5
1000	8.4
1500	7.5
2000	6.6
2500	5.9
3000	0.7

If the cylinder is not leaking internally, then it must be the pump, right? To prove this assumption, we relocate the flow-tester so that it is positioned after the pump, but before the relief valve, directional control valve and cylinder.

A word of warning here - when a flow-tester is connected upstream of the relief valve, in a circuit with a fixed-displacement pump, care must be taken not to over-pressurize the pump through overzealous use of the load valve. Most flow-testers are fitted with a pressure limiting device, however this may be set much higher than the pressure rating of the pump that is being tested.

The results of the pump test look like those shown in exhibit 2.4. As you can see, pump flow is 9.8 GPM at 50 PSI and 9.0 GPM at 3000 PSI. This represents a volumetric efficiency of 90% at 3000 PSI ($9.0 \div 10 \times 100 = 90$). This tells us that the pump is not the cause of the problem.

Exhibit 2.4

PRESSURE PSI	FLOW RATE GPM
50	9.8
500	9.8
1000	9.7
1500	9.5
2000	9.4
2500	9.3
3000	9.0

In the first test, flow decreased significantly at 3000 PSI (see exhibit 2.3) because the relief valve opened, allowing the remaining flow to bypass the flow-tester. In the second test, the flow-tester was connected into the circuit upstream of the relief valve and so the relief valve had no influence on the flow readings.

We have now established that neither the cylinder nor the pump is causing the problem. It must be the relief valve that is leaking internally, right? To prove this assumption, we relocate the flow-tester so that it is positioned after the pump and relief valve but before the directional control valve and cylinder.

The results of this test look like those shown in exhibit 2.5. As you can see, apart from the influence of the relief valve at 3000 PSI, when the flow decreases significantly due to the relief valve opening and allowing the remaining flow to bypass the flow-tester, the results of this test are identical to the previous test. This indicates that the relief valve is not leaking internally.

Exhibit 2.5

PRESSURE PSI	FLOW RATE GPM
50	9.8
500	9.8
1000	9.7
1500	9.5
2000	9.4
2500	9.3
3000	0.5

As a result of this process of elimination, we have now isolated the problem to the directional control valve. This is confirmed when disassembly of the directional control valve reveals a crack in the casting, which is allowing fluid to pass from the pressure gallery to the tank gallery.

This hypothetical example illustrates how a flow-tester is used to locate internal leakage in a hydraulic system. It also demonstrates how easy it can be to jump to the wrong conclusions in a troubleshooting situation. This leads to incorrect diagnosis of the problem, which usually results in the unnecessary repair or replacement of serviceable components. To avoid such costly mistakes, the correct equipment and a logical approach are required. For more information on flow testing equipment, go to www.HydraulicSupermarket.com/flowtest

10

Troubleshooting hydrostatic transmissions

The troubleshooting principles described in the previous two Chapters can be applied in both open and closed circuits. However, troubleshooting closed-circuit, hydrostatic transmissions requires additional knowledge and specialized techniques. Before I discuss these in detail, let me explain the operating principles of closed-circuit transmissions.

HYDROSTATIC TRANSMISSION OPERATING PRINCIPLES

A hydrostatic transmission consists of a variable-displacement pump and a fixed or variable displacement motor, operating together in a closed circuit. In a closed circuit, fluid from the motor outlet flows directly to the pump inlet, without returning to the reservoir. A simple closed circuit is illustrated in exhibit 2.6.

Exhibit 2.6

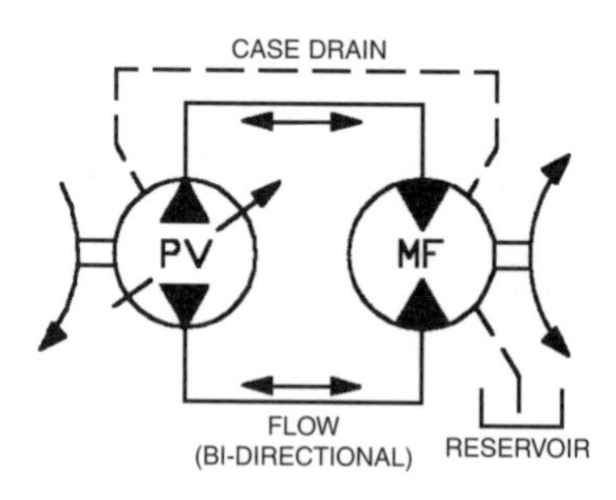

Basic closed circuit comprising
variable displacement pump (PV)
and fixed displacement motor (MF).

As well as being variable, the output of the transmission pump can be reversed, so that both the direction and speed of motor rotation are controlled from within the pump. This eliminates the need for directional or flow (speed) control valves in the circuit. The pump and motor may be integral, where both are contained in a single assembly; or split, where each is a separate unit, connected together by pipes or hoses.

The simple, closed circuit (loop) illustrated in exhibit 2.6 will not operate effectively in practice. This is because the pump and motor leak internally, which allows fluid to escape from the loop and drain back to the reservoir. To compensate for these losses, a fixed-displacement pump called a charge pump is used, to ensure the loop remains full of fluid during normal operation. The charge pump is normally installed on the back of the transmission pump and has an output of at least 20% of the transmission pump's output.

In practice, the charge pump not only keeps the loop full of fluid; it pressurizes the loop to between 110 and 360 PSI, depending on the transmission manufacturer. A simple charge pressure circuit comprises the charge pump, a relief valve and two check valves, through which the charge pump can replenish the transmission loop. Once the loop is charged to the pressure setting of the relief valve, the flow from the charge pump passes over the relief valve, through the case of the pump or motor or both, and back to the reservoir. This is illustrated in exhibit 2.7.

Exhibit 2.7

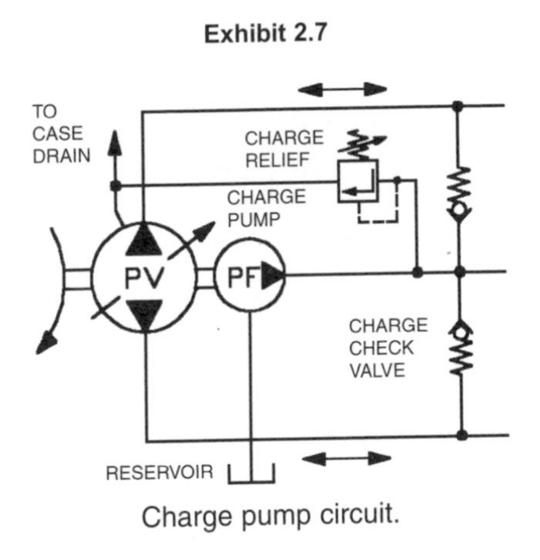

Charge pump circuit.

To protect the transmission pump and motor from damage through over-pressurization, high-pressure relief valves are included in the circuit. This is illustrated in exhibit 2.8. The charge relief and check valves, and the high-pressure relief valves are usually built into the transmission pump. However, the charge and high-pressure relief valves are sometimes built into the transmission motor.

Exhibit 2.8

CHARGE
CHECK
VALVE
PV
HIGH
PRESSURE
RELIEFS
FROM
CHARGE
PUMP

High pressure relief valve circuit.

Hydrostatic transmissions used to propel mobile machinery are often fitted with a bypass valve, which allows the machine to be towed. When activated, the bypass valve allows fluid to pass across the loop. This permits the hydraulic motor to rotate freely when it is driven by the wheels, when the machine is towed. The arrangement of a bypass valve is illustrated in exhibit 2.9. It is important that the bypass valve is fully closed during normal operation, otherwise the pressure drop across the valve will cause the transmission to overheat.

Exhibit 2.9

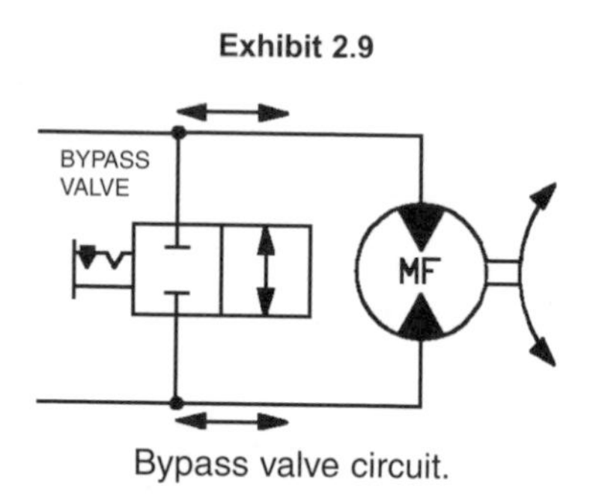

Bypass valve circuit.

CHECKING CHARGE PRESSURE

Charge pressure is a good indicator of the condition of a hydrostatic transmission and is one of the first things that should be checked when troubleshooting. In order to check charge pressure, a 0-900 PSI pressure gauge must be connected to the charge pressure gauge port on the transmission pump. The location of this gauge port varies, therefore consult the machine manufacturers' instructions. Once located, fitting a test-point will enable charge pressure to be easily checked at any time. For more information on pressure testing equipment, go to www.HydraulicSupermarket.com/pressuretest

A transmission in good condition will hold its specified charge pressure in neutral, and this value should not decrease by more than 10% when the transmission is operating in forward or reverse. If charge pressure is low in neutral or drops significantly in forward or reverse, this indicates either a problem with the charge pump circuit or that the amount of leakage from the transmission loop is approaching the capacity of the charge pump. The latter is usually a result of abnormal leakage from the transmission pump and/or motor due to internal wear or damage.

CAVITATION AND ITS CAUSES IN HYDROSTATIC TRANSMISSIONS

Low charge pressure can result in cavitation. Cavitation occurs when the volume of fluid returning from the outlet of the motor to the inlet of the transmission pump, combined with the volume of fluid being delivered by the charge pump, is less than the volume of fluid being demanded by the transmission pump. This causes the fluid at the pump inlet to vaporize, which results in the formation of cavities within the fluid. These cavities implode when compressed on the outlet side of the transmission pump.

Common causes of cavitation in hydrostatic transmissions include:

- clogged charge pump inlet filter;
- restricted charge pump intake line;
- worn or damaged charge pump;
- excessive internal leakage (worn or damaged transmission pump and/or motor); and
- excessive external leakage (burst hose or pipe).

Another cause of cavitation in hydrostatic transmissions which is frequently overlooked, is the combined effect of fluid compressibility and the 'accumulator effect' of conductors (the increase in volume of a hose or pipe as pressure increases).

When a hydrostatic transmission is subject to a sudden increase in load, the motor stalls instantaneously and system pressure increases until the increased load is overcome or the high pressure relief valve opens - whichever occurs first.

While the motor is stalled, there is no return flow from the outlet of the motor to the inlet of the pump. This means that the transmission pump will cavitate for as long as it takes to make-up the volume of fluid required to develop the pressure needed to overcome either the increased load or the high-pressure relief valve. How long the pump cavitates depends on the output of the charge pump, the magnitude of the pressure increase, and its influence on the increase in volume of the conductor and the decrease in volume of the fluid. This is illustrated in the following example.

A hydrostatic transmission operating the drill head on a drill rig is delivering a flow of 35 GPM at a pressure of 1000 PSI. A sudden increase in load on the drill bit instantaneously stalls the motor until sufficient pressure is developed to overcome the increase in load, which for the purposes of this example is 3000 PSI.

In order to increase system pressure from 1000 PSI to 3000 PSI, the transmission pump must make-up additional volume, due to the compression of the hydraulic fluid and the volumetric expansion of the high-pressure hose between the pump and the motor. But because the motor is momentarily stalled, there is no return flow from the outlet of the motor to the inlet of the pump. The only fluid available at the inlet of the transmission pump is 7 GPM from the charge pump, which is around 80% less than required!

In this example, the high-pressure hose between the pump and motor is SAE 100R9AT-16, 36 feet long. The volumetric expansion of this hose, due to the increase in pressure, is 9.7 in^3 and the additional volume required due to compression of the fluid within this hose is 2.8 in^3. Therefore the total, additional fluid volume required to increase the operating pressure from 1000 to 3000 PSI is 12.5 in^3 (9.7 + 2.8 = 12.5).

To calculate the time taken for the operating pressure to increase from 1000 to 3000 PSI, which is equivalent to the length of time the transmission pump will cavitate, we divide the required make-up volume (12.5 in^3) by the volume available from the charge pump per second (27 in^3). In this example, the transmission pump cavitates for 0.46 seconds every time a sudden increase in load demands an increase in system pressure from 1000 to 3000 PSI ($12.5 \div 27 = 0.46$).

This problem occurs in applications where there is insufficient charge volume installed to cope with the effects of sudden fluctuations in load on the transmission. Typical examples include drill rigs, boring machines, and cutter wheels on dredgers. In applications where this problem occurs, additional charge volume will need to be installed. Consult the machine manufacturer or a fluid power engineer for guidance.

Cavitation increases heat load and causes metal erosion, which results in damage to the pump and contamination of the hydraulic fluid. In extreme cases, cavitation can cause mechanical damage to the transmission, resulting in major failure. For this reason, NEVER operate a hydrostatic transmission with low charge pressure.

HIGH FLUID TEMPERATURE

When troubleshooting a hydrostatic transmission, as with any hydraulic system, always begin by checking the obvious things first. Check the fluid level in the reservoir and, if low, fill to the correct level. Check that there are no obstructions to airflow around the reservoir, such as a build up of dirt or debris. Inspect the heat exchanger and ensure that the core is not blocked. Check the efficiency of all cooling circuit components and replace as necessary.

Remember that aeration and cavitation increase heat load. Inspect the charge pump intake for possible causes of aeration and cavitation, particularly if the transmission is noisy.

Check that the bypass valve, if fitted, is fully closed. Leakage across the bypass valve creates a pressure drop that increases the heat load on the transmission.

Check charge pressure. If charge pressure is normal, then the transmission may be overloaded. Overloading of the transmission can cause system pressure to reach or exceed the pressure setting of the relief valve. Leakage across the high-pressure relief valve creates a pressure drop, which increases the heat load on the transmission. If this is the case, it will be necessary to either reduce the load on the transmission or increase the pressure setting of the relief valve(s). **A word of warning here** - do not increase the pressure setting of the relief valves without consulting the machine manufacturer.

If charge pressure is low, this is an indication that the high fluid temperature is being caused by abnormal internal leakage from the pump, motor or both. This can be confirmed by conducting flow tests on the transmission. Even though abnormal internal leakage may be isolated to either the pump or the motor, both components should be removed for inspection and repair. The logic for this is quite simple. The pump and motor work together under identical operating conditions and therefore, if one is in need of repair, it is safe to assume that the other will soon be in need of the same attention.

If the pump is removed for rebuild and the motor is not, then the likely outcome is that the motor will fail subsequently, causing damage to the recently rebuilt pump. Remember that in a closed circuit, fluid flows directly from the outlet of the motor to the inlet of the pump, usually without being filtered. This means that if the motor starts to fail, generating metal particles in the process, this contamination will flow directly to the pump, causing damage (and vice versa). It is false economy to rebuild one component of a hydrostatic transmission and not the other.

SLOW OPERATION

As with any hydraulic system, a loss of speed from a hydrostatic transmission indicates a loss of flow. Check the bypass valve (if fitted) is fully closed. Check charge pressure. As already explained, low charge pressure can indicate that the volume of internal leakage from the transmission loop is approaching the capacity of the charge pump. This can be confirmed by conducting flow tests.

Low charge pressure can also prevent the transmission pump from maintaining full displacement. This is because in most transmissions, charge pressure is used to control the displacement of the pump. If there is insufficient charge pressure to hold the pump at full displacement under load, pump flow, and therefore motor speed, will decrease as the load on the transmission increases. This problem is easy to identify on transmission pumps fitted with an external, displacement angle indicator.

Note that this type of problem does not involve a pressure drop or its associated heat generation. This means that if a transmission is slow in operation because of a displacement control problem, fluid temperature should be normal. In this case, the low charge pressure is likely to be the result of a problem in the charge circuit.

Check that the charge pump inlet filter, if fitted, is not clogged and the intake line to the charge pump is not restricted. If charge pressure is still low, the charge pump may be worn or damaged, in which case the transmission should be removed for inspection and repair.

11

How to avoid costly troubleshooting mistakes and rip-offs

In this and following Chapters, I describe various scams and rip-offs perpetrated by some elements of the fluid power industry. While it is difficult to reveal these dishonest techniques without appearing negative, I am certainly not suggesting that the entire industry is crooked. But there are unethical operators out there who will take advantage of you. The good news is that if you know the pitfalls, you can avoid them.

The previous four Chapters are not intended to make you an expert in hydraulic troubleshooting. However, when you have a problem with your hydraulic equipment, this information can save you a lot of time and money. In this Chapter, I show you how.

ASSESSING THE PROBLEM AND ELIMINATING THE OBVIOUS

Before you incur the expense of hiring a technician, make sure you have assessed the problem and eliminated all of the obvious, possible causes. I have lost count of the number of times that I've been called to a problem and found that the cause was something quite simple. A wire broken off a solenoid valve, a pin fallen out of a mechanical linkage, an isolation valve that had vibrated closed, a blocked heat exchanger... and so the list goes on.

This won't bother the technician, because his hourly rate is the same, regardless of how easy or difficult the problem is to find. However, you may be annoyed with yourself for not checking something so obvious, knowing that you could have easily saved yourself a couple of hundred dollars.

After you have assessed the problem and eliminated the obvious, possible causes, you can continue the process of elimination by carrying out additional checks, assuming you have the necessary test equipment and are familiar with the circuit in question. A set of pressure gauges, an infrared heat gun and a photocell tachometer can be used to locate most problems.

More specialized test equipment such as flow-testers, peak-pressure meters, and multi-meters, will be required to carry out checks on complex systems or systems with electric or electronic controls. For more information on hydraulic test equipment, go to www.HydraulicSupermarket.com/hydtest

CHOOSING WHERE TO GO FOR SPECIALIST HELP

Let's assume you have assessed the problem, checked all the obvious, possible causes and eliminated what you can with the test equipment available. We have now established that we are going to need specialist help. Who do you call? There are normally three options: the machine dealer; the manufacturer of the hydraulic components; or an independently-owned hydraulic shop.

In theory, the machine dealer's service technicians should be the best, due to their machine-specific training provided by the factory. However, because these technicians are normally expected to service all of the various systems on the machine, they are typically all-rounders rather than hydraulic specialists. The machine dealer usually charges a premium for their technician's factory training and state-of-the-art equipment.

It is common for the majority of components that comprise a machine's hydraulic system to be supplied by one particular manufacturer. Therefore, if the hydraulic system has a problem, it is logical to call the manufacturer of the hydraulic components. Note here that I am referring to a service center that is owned and operated by the hydraulic component manufacturer, rather than an independent hydraulic shop that is an authorized service center and/or distributor of the manufacturer's products.

In addition to being hydraulic specialists, the component manufacturer's service technicians have the advantage of factory training and access to the manufacturer's service literature. And like the machine dealer, hydraulic component manufacturers usually charge a premium for their technicians.

Technicians employed by independently-owned hydraulic shops don't usually have access to the same level of product-specific training or technical literature as technicians who work for the hydraulic component manufacturers. This tends to apply even if the independent is an authorized service center and/or distributor of the manufacturer's products. However, like most people, hydraulic technicians change jobs and many technicians employed by independent hydraulic shops have gained experience working for hydraulic component manufacturers. Independents don't usually charge premium rates for their technicians like machine dealers and hydraulic component manufacturers do.

WHY QUALITY IS MORE IMPORTANT THAN QUANTITY

My experience has been that mediocre technicians outnumber the good ones, and unfortunately it is not possible to determine a technician's competency from the badge on his shirt or his charge-out rate. While charge-out rates may be a factor in deciding whose technician you hire, from an overall cost perspective it is far more important to evaluate the technician and his diagnosis, so that you don't end up paying for his mistakes or fall victim to a troubleshooting rip-off.

Let me illustrate how easily this can happen with an example. Several years ago, I was asked for a second opinion on the condition of a set of pumps operating a processing plant. The customer had called in a technician to check the performance of these pumps and was alarmed when the technician advised that all four pumps were in need of repair.

The pumps in question were variable-displacement units fitted with constant power control. The power required to drive a hydraulic pump is a product of flow and pressure. A constant power or power limiting control operates by reducing the displacement, and therefore flow, from the pump as pressure increases, so that the power rating of the prime mover is not exceeded. The advantage of this type of control is that more flow is available at lower pressures, so that the machine operates faster under light loads. This results in better utilization of the power available from the prime mover.

Pump performance is checked using a flow-tester to load the pump and measure its flow rate. As resistance to flow is increased, pressure increases and the flow available from the pump to do useful work decreases because of internal leakage. The difference in the measured flow rate between no load and full load determines the volume of internal leakage and therefore pump performance.

I tested all four pumps, recording flow against pressure from no load through to maximum working pressure. In my report I explained to the customer that the tests revealed that pump flow did decrease significantly as pressure increased, but that this is a normal characteristic of a pump fitted with constant power control. I further advised that apart from the constant power control requiring adjustment on two of the pumps, the performance of all four pumps was acceptable.

The first technician's assessment can only be explained by fraud or incompetence. I suspect it was the latter, with the technician failing to either establish or understand that the pumps he was testing were fitted with constant power control. This ignorance led to an incorrect interpretation of the test results. Whatever the explanation, the customer could have paid thousands of dollars for unnecessary repairs, if they had not sought a second opinion.

ASSESSING THE COMPETENCY OF A HYDRAULIC TECHNICIAN

Let's assume you have chosen a service provider and their technician has arrived on site. How do you know if he is a good one? There are a few things to look for when evaluating a hydraulic technician.

After you have described the nature of the problem, a competent technician should do two things before he reaches for his test equipment. The first is to conduct a visual inspection of the hydraulic system, checking all the obvious things that could cause the problem you have described (he will not be aware that you have done this already). The second is to ask for a schematic diagram for the hydraulic circuit. The only possible exceptions would be where the technician is already familiar with your particular hydraulic machine or the circuit is extremely simple.

A hydraulic schematic diagram is a 'road map' of the hydraulic system, which is a valuable aid in identifying possible causes of a problem. A schematic diagram can save a lot of time in a troubleshooting situation and it is reasonable to expect that a hydraulic technician is able to read and interpret one.

Having said that, a competent technician should be equally capable of working without a schematic diagram, but this involves tracing the physical circuit and identifying its components in order to isolate possible causes of the problem.This can be a time-consuming process, depending on the complexity of the system.

TROUBLESHOOTING RIP-OFFS

Armed with the correct equipment, a competent technician should be able to identify the majority of problems, on all but the most complex systems, in less than two hours. If the technician appears to be proceeding in an illogical manner, is making inconsistent statements or using a lot of technical jargon, one of two things is probably happening.

Either the technician is not competent and you are paying for his on-the-job training, or he is perpetrating a scam that I call 'stretching it out'. In this scam, the technician establishes the cause of the problem quite quickly, but instead of informing you and starting to fix it, he continues to look in another part of the circuit. Once a suitable amount of time has been wasted, he then 'discovers' the problem. By doing this, the technician intentionally prolongs or 'stretches out' the service call.

In either situation, you are being ripped-off. If you have any doubts about the competency or integrity of the technician working on your machine, call the company's service supervisor. Explain that you think the technician may be out of his depth and request a more experienced technician. The company will most likely respond by dispatching their best technician to fix your problem - as quickly as possible.

GETTING THE CORRECT DIAGNOSIS – THE FIRST TIME

Paying for more of a technician's time than is required is certainly not desirable. But it is nowhere near as costly as paying for the unnecessary repair or replacement of serviceable components, as a result of incorrect diagnosis of a problem. Incorrect diagnosis in a troubleshooting situation is usually a result of the technician's incompetence, insufficient investigation of the problem or a combination of both.

To illustrate this point, consider the hypothetical example used in Chapter 9, which explains how a flow-tester is used to isolate internal leakage. The simple hydraulic circuit in this example comprised only four components of interest, a pump, a relief valve, a directional control valve and a double-acting cylinder. The cylinder was slowing under load, the cause of which turned out to be a crack in the body of the directional control valve.

Let's say you called in a technician to find the problem. The technician arrives and immediately notices how simple the system is. You tell him that you had the seals in the cylinder replaced not so long ago. This statement adds weight to the technician's immediate assumption that the pump must be the cause of the problem. But he gets you to operate the system anyway. He looks and listens intently while the system operates. This exercise confirms his diagnosis and he advises you that the pump is worn-out and needs to be repaired. You accept the technician's incorrect diagnosis without question, so he removes the pump from the machine and takes it away for repair.

After several days of downtime, the technician returns with the now repaired pump and installs it on the machine. As soon as the machine is operated, it is obvious that rebuilding the pump has not solved the problem. The now embarrassed technician gets out his flow-tester and discovers, through a logical process of elimination, that the cause the problem is the cracked directional control valve.

This is an example of an incorrect diagnosis resulting from insufficient investigation of the problem, rather than incompetence, although there is a fine line between the two.

Unless the pump was in new condition, it would not be difficult for the technician to demonstrate that the repairs were justified, and therefore you end up paying for repairs unrelated to the cause of the original problem. It is important to understand that this unnecessary repair could have been avoided, if the technician had been asked to prove that the pump was worn-out using a flow-tester before the pump was removed from the machine.

Although I encourage you to assess the problem and check the obvious, possible causes before you call for specialist help, don't fall into the trap of sharing your diagnosis with the technician.

Consider another scenario on the same system. This time when the technician arrives on site, you advise him that the cylinder appears to be generating heat and, as far as you know, its seals have never been replaced. The technician will probably be happy to go along with your assessment of the problem and remove the cylinder for repair. Of course, this story ends the same way as the last one. You end up paying for repairs that are unrelated to the actual problem.

To avoid this trap, allow the technician to develop his own diagnosis of the problem before you tell him what you think it might be. Forming your own idea of the problem and keeping it to yourself puts you in a good position to determine if the technician's explanation makes sense.

The most important thing however, is to ensure that whenever a component is identified as needing repair or replacement, the diagnosis is based on a logical process of elimination using the appropriate test equipment and not on intuition or guesswork.

PART II CONCLUSION

Gaining a solid understanding of the theory and principles of troubleshooting described in the preceding Chapters, and applying them in practice, may require some time and effort. But this small investment in self-education will save you thousands of dollars in unnecessary service call-outs, repairs and machine downtime.

PART III

"Let the buyer beware" Insider Secrets to Repairing & Replacing Hydraulic Components

12. Fundamentals of hydraulic component repair
13. Repairing hydraulic cylinders
14. Choosing a hydraulic repair shop
15. Evaluating a hydraulic repair shop
16. Common repair rip-offs and how to avoid them
17. How to buy replacement components at the lowest possible price
18. How to safeguard new component warranty and get free repairs after warranty has expired

INTRODUCTION

All hydraulic components have a finite service life and at some point will require repair or replacement. The objective of the following Chapters, is to explain the fundamentals of hydraulic component repair and how repair shops and fluid power distributors operate, so that when you do need to replace a hydraulic component, you can be sure of three things. Firstly, if the component is repairable, you will have some understanding of the type of repair required and an idea of how it can be undertaken in the most economical way. Secondly, if you proceed with a repair, you will be able to recognize and avoid common repair rip-offs. And finally, if you need to buy a new component, you will know how to buy it at the lowest possible price.

12

Fundamentals of hydraulic component repair

Rebuilding a hydraulic component involves reworking or replacing all of the parts necessary to return the component to 'as new' condition, in terms of performance and expected service life. This means that a rebuilt component should perform as well and last as long as a new one. In this Chapter, I will use the terms 'rebuild' and 'repair' interchangeably with the same meaning, as defined above.

In many cases, rebuilding a component can result in significant savings when compared with the cost of purchasing a new one. The economics of proceeding with any repair is ultimately dependent on the cost of the repair, relative to the cost of a new component. As a rule, the more expensive a new component is in absolute dollar terms, the more likely it is that a repair will be cost effective.

FACTORS THAT INFLUENCE COMPONENT REPAIR COSTS

A component's rebuild cost is determined by a number of factors including:

- extent of wear or damage to the component;
- facilities and expertise of the repairer; and
- repair techniques employed.

As explained in Chapter 4, the first step in minimizing component repair costs is to change-out components in pre-failed condition upon completion of their expected service life. A failure in service usually results in mechanical damage to the internal parts of the component and as a consequence, parts that would have been serviceable, have to be replaced.

While some repairs to hydraulic components can be carried out effectively by personnel who are competent in mechanical fitting and machining, many of the techniques used to rebuild hydraulic components require specialist knowledge and equipment. The facilities and expertise of the repairer and the techniques they employ can have a significant influence on the viability of a repair, as the following example illustrates.

Some years ago, when I was working for a specialist, hydraulic rebuild shop, a customer asked us to quote on rebuilding a large spool-type directional control valve. The valve in question had been badly damaged as a result of cavitation, which had occurred over a long period in service.The metal erosion in the body of the valve was so severe that two other repair shops had advised the customer that the valve was not repairable.

A technique commonly used to repair this type of valve involves machining the spool bores oversize to remove scoring or cavitation damage, and manufacturing oversize spools to match the oversize bores.The problem with this particular valve was that the damage was too deep for this method of repair to be successful. But the price of a new valve was $20,000 and the customer was determined to explore all possible options before purchasing a new one.

In response to the customer's situation, we developed a repair technique for this valve, which involved:

- machining the spool bores oversize until all cavitation damage was removed;
- manufacturing sleeves which were externally threaded and screwed into the valve body;
- machining the sleeved bores oversize; and
- manufacturing oversize spools to match.

This was a complicated repair that required considerable expertise and specialist facilities. It was also a relatively expensive repair that was only economic due to the high replacement cost of a new valve. But because the cost to repair the valve was a fraction of new price, the customer saved a considerable amount of money.

DIY REPAIRS

Not all repairs to hydraulic components are as complicated as the one described in the above example. The extent to which hydraulic component repairs can be undertaken in-house depends on the type of component and its condition, the expertise of the personnel carrying out the repairs and the facilities and information available to them. For example, there is a significant difference in the equipment and expertise required to replace the seals in a cylinder, compared with that required to completely rebuild a variable-displacement piston pump.

If you do carry out repairs to hydraulic components yourself, always follow the component manufacturer's repair instructions. Unfortunately, this information can be difficult to obtain. Most hydraulic component manufacturers treat their repair instructions as valuable, intellectual property and do not circulate them freely.

The best way to obtain repair instructions for a machine's hydraulic components is at the time of purchase, by making it a condition of the contract. The machine manufacturer usually has the necessary influence to obtain repair instructions from the hydraulic component manufacturers and should be willing to use this influence, if the sale of a machine depends on it.

13

Repairing hydraulic cylinders

As a product group, cylinders are almost as common as pumps and motors combined. They are less complicated than other types of hydraulic components and are therefore relatively easy to repair. Like all hydraulic components, cylinders have a finite service life and at some point will need to be rebuilt or at least re-sealed.

The following is a general guide to repairing hydraulic cylinders. The extent of the repair work that can be carried out in-house depends on the extent of wear or damage to the cylinder and how well equipped your repair shop is. As with any repair, the economics of proceeding with a repair on a cylinder are ultimately dependent on the cost and availability of a new one.

DISASSEMBLY AND INSPECTION

Typically, a cylinder will have been removed for repair due to either external or internal leakage. Close inspection of the parts of the cylinder after disassembly, particularly the seals, can reveal problems that may not otherwise be obvious.

Piston seal - If the piston seal is badly distorted, eroded or missing completely, this indicates that the barrel is oversize or has bulged in service. In this case, the barrel or the complete cylinder should be replaced. Replacing the piston seal without replacing the barrel is a short-term fix only.

Rod seal - If the rod seal is badly distorted, this usually indicates that either the guide bush is excessively worn or the rod is bent. In both cases this results in the weight of the rod riding on the seal, which causes it to fail. Replacing the rod seal without identifying and rectifying the cause of the problem is a short-term fix only.

Rod - Check the rod for cracks at all points where its cross-section changes. Dye penetrant is ideal for this purpose. It is easy to use and readily available from industrial hardware merchants.

Inspect the chrome surface of the rod. If the chrome looks dull on one side and polished on the opposite side, this indicates that the rod is bent. Rod straightness should always be checked when a cylinder is being repaired. This is done by placing the rod on rollers and measuring the run-out with a dial gauge (see exhibit 3.1). Position the rod so that the distance between the rollers (L) is as large as possible and measure the run-out at the mid-point between the rollers (L/2).

Exhibit 3.1

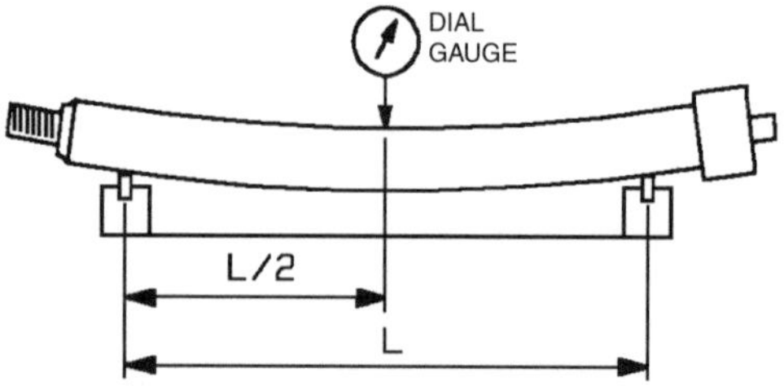

Checking rod straightness.

The rod should be as straight as possible, but a run-out of 0.5 mm per linear meter of rod is generally considered acceptable. To calculate maximum, permissible run-out (measured at L/2) use the formula:

Run-out max. (mm) $= 0.5 \times L \div 1000$

Where L equals distance between rollers in millimeters.

For example, if the distance between the rollers was 1.2 meters, then the maximum, allowable run-out measured at L/2 would be given by $0.5 \times 1200 \div 1000 = 0.6$mm.

In most cases, bent rods can be straightened in a press. It is sometimes possible to straighten rods without damaging the hard-chrome plating, however if the chrome is damaged, the rod must be either re-chromed or replaced.

If the chrome surface of the rod is pitted or scored, the effectiveness and service life of the rod seals will be reduced. Minor scratches in the chrome surface can be polished out using a strip of fine emery paper in a crosshatch action. If the chrome is badly pitted or scored, the rod must be either re-chromed or replaced. Machining a new rod from hard-chrome plated round bar, which is available in standard sizes from specialist steel merchants, is usually the most economical solution for small diameter rods.

Before a rod can be re-chromed, the existing chrome plating has to be ground off. Each time a rod is ground, the diameter of the parent metal is reduced and therefore the thickness of the chrome layer required to finish the rod to its specified diameter increases. If the chrome layer is too thick, the chrome will stress crack, resulting in premature failure of the rod seals. Therefore, when the thickness of the chrome plating on a cylinder rod reaches 0.008" the rod must be scrapped.

Chrome thickness can be measured using a coating-thickness gauge. A coating-thickness gauge uses the magnetic properties of the substrate material to determine the thickness of a nonmagnetic coating. For more information on instruments for measuring chrome thickness, go to www.HydraulicSupermarket.com/chrometest

Head - It is common, in cylinders used in light-duty applications, for the rod to be supported directly on the head material, which is usually aluminium alloy or cast iron. A metallic or non-metallic guide bush (wear band) is fitted between the rod and the head, in applications where there are high loads on the rod. If a cylinder is fitted with a bush between the rod and the head, it should be replaced as part of the repair.

If the rod is supported directly on the head, use an internal micrometer or vernier calliper to measure the head's internal diameter. Take measurements in two positions, 90 degrees apart, to check for ovalness. The inside diameter of the head should not exceed the nominal rod diameter plus 0.004". For example, if the nominal diameter of the rod is 1-1/2" then the inside diameter of the head should not exceed 1.504". If the head measures outside this tolerance, it will allow the rod to load the rod seal, resulting in premature failure of the seal. Therefore, the head must be sleeved using a bronze bush or be replaced with a new head, machined from a similar material.

Minor scoring on the lands of the seal grooves inside the head is not detrimental to the function of the cylinder, as long as the maximum diameter across the lands does not exceed the nominal rod diameter plus 0.016". For example, if the nominal diameter of the rod is 1-1/2" then the inside diameter of the head, measured across the lands of the seal grooves, should not exceed 1.516". If the seal lands measure outside this tolerance, the service life of the rod seal will be reduced. Therefore, the head must be replaced with a new head, machined from a similar material.

Barrel - Inspect the barrel for internal pitting or scoring. If the barrel is pitted or scored, the effectiveness and service life of the piston seal will be reduced. Therefore, the barrel must be honed to remove damage or be replaced. On small diameter barrels, pitting or scoring less than 0.005" deep can be removed using an engine-cylinder honing tool. The barrel must be honed evenly along its full length.

The maximum bore diameter for standard-size piston seals is the nominal bore diameter plus 0.010". For example, if the nominal bore diameter of the barrel is 2-1/2" then the maximum size after honing should not exceed 2.510". This size should be checked at several points along the barrel, using an internal micrometer.

If scoring or pitting is still present at 0.010" oversize, the barrel must be honed further to accommodate oversize seals or be replaced. Manufacturing a new barrel from honed tubing, which is available in standard sizes from specialist steel merchants, is usually the most economical solution for small diameter cylinders.

Large diameter, inch-size cylinder barrels can be salvaged by honing either 0.030" or 0.060" oversize and fitting the corresponding oversize piston seals. Oversize seals for metric-size cylinders have limited availability and therefore it is not always possible to salvage metric-size barrels by fitting oversize seals.

Piston - The pistons of cylinders used in light-duty applications are usually machined from aluminium alloy or cast iron and operate in direct contact with the cylinder bore. Minor scoring on the outside diameter of the piston is not detrimental to the function of the cylinder, as long as the minimum diameter of the piston is not less than the nominal bore diameter minus 0.006". This can be checked using an external micrometer. For example, if the nominal diameter of the barrel is 2-1/2" then the minimum piston diameter would be 2.494". If the piston diameter measures outside this tolerance, it must be replaced with a new piston, machined from a similar material.

Non-metallic wear bands are fitted between the piston and barrel, in applications where there are high loads on the rod. If the cylinder is fitted with piston wear bands, these should be replaced as part of the repair.

ORDERING SEALS

If you order seals from a seal supplier, avoid the common practice of measuring the old seals. Seals can either shrink or swell in service and in some cases, an incorrect seal may have been installed previously. To ensure that you are supplied with the correct seals, measure all seal grooves with a vernier calliper and give this information to your seal supplier. For more information on hydraulic seals, go to www.HydraulicSupermarket.com/seals

ASSEMBLY

Thoroughly clean all parts in a petroleum-based solvent and blow-dry using compressed air. Coat all parts with clean hydraulic fluid during assembly. Prior to installing seals, ensure that the seal grooves are clean and free from nicks and burrs. Avoid using a screwdriver or other sharp object when installing seals, as this can result in damage to the seal. After the cylinder has been assembled, plug its service ports to prevent ingress of moisture or dirt.

14

Choosing a hydraulic repair shop

If you are responsible for keeping one or more hydraulic machines running and are serious about minimizing your operating costs, then at some point, you will need to have a hydraulic component rebuilt by a specialist.

There are normally three options: the machine dealer; the component manufacturer; or an independently-owned hydraulic shop. As explained in Chapter 12, the facilities and expertise of the repairer and the repair techniques they employ can have a significant impact on the price you pay to have a component rebuilt.

MACHINE DEALERS

When it comes to hydraulic component repairs, the capabilities of machine dealers can vary widely. Some have the necessary expertise and equipment to carry out most hydraulic rebuilds in-house, while others are totally reliant on outside suppliers to provide this service. You will usually pay more to have a hydraulic component rebuilt by the machine dealer, particularly if they rely on a third party to carry out the repair.

EXCHANGE PROGRAMS

If machine downtime is an issue and the dealer offers an exchange or 'reman' program, this can be an attractive option. The benefit of an exchange program is that the dealer carries inventory of rebuilt components. So instead of having to wait for your component to be repaired, you can exchange your component with a reman unit from the dealer's stock and pay the dealer's exchange price.

It is important to understand that the exchange price is not necessarily based on what it would cost to repair your particular component. The exchange price is normally based on the average rebuild cost for that component plus an inventory charge, and may be conditional on certain parts of your component being reusable.

Because the dealer's exchange price is based on an average rebuild price, if you were to have your component repaired it could cost more than the exchange price, but it could also cost a lot less! Of course, machine downtime costs money and the main advantage of an exchange program is that downtime is minimized.

If downtime is not an issue, then it is wise to get a quote to repair your component before opting for an exchange unit. This is another reason for scheduling component change-outs upon completion of their expected service life. If component rebuilds can be scheduled when machine downtime is not an issue, it will not be necessary to pay a premium for an exchange unit.

HYDRAULIC COMPONENT MANUFACTURERS

When a hydraulic component needs repairing, it may seem logical to send it to the company that manufactured it in the first place. Note here that I am referring to a repair shop that is owned and operated by the hydraulic component manufacturer, rather than an independent hydraulic shop that is an authorized repairer and/or distributor of the manufacturer's products. There are a couple of traps to watch out for, if you send a hydraulic component to the manufacturer for repair.

Firstly, hydraulic component manufacturers are almost always more interested in selling you a new component than they are in fixing your old one. This makes sense when you consider that they are in business to make and sell as many new hydraulic components as possible. Consequently, they may use various techniques to get you to buy a new component. They may tell you that your old component is uneconomic to repair or that it is has been superseded by a new model and therefore your particular model is no longer supported, even though your machine may only be a couple of years old. Whatever the story, their solution will be for you to buy a new component, which removes the opportunity for you to save money by having your old component rebuilt.

Secondly, if a hydraulic component manufacturer can't sell a new component, then the next best thing as far as they are concerned is to sell as many spare parts as possible. This means that when they do rebuild a component, they are not interested in minimizing the cost of the repair by employing money-saving repair techniques. This is best illustrated with an example.

Let's say you have removed a radial piston motor from your hydraulic machine and shipped it to the motor manufacturer for repair. When the motor is dismantled, it is discovered that several of the pistons and their bores are badly scored as a result of contamination. The manufacturer advises you that because of this damage, the motor's housing and all of its pistons need to be replaced. As a result, the manufacturer's price to repair this motor is almost as much as a new one.

You don't want to buy a new motor if you can avoid it, so you decide to get a second repair quote. This time you take the unit to a reputable, independently-owned, hydraulic shop. After inspecting your motor, this repair shop advises you that they can rebuild it for around 60% of the cost of new one - a significant saving on a $10,000 motor!

They are able to do this by employing a money-saving repair technique which involves machining the piston bores oversize to remove scoring, and fitting a set of oversize pistons, thereby salvaging the housing. The oversize pistons are non-genuine or aftermarket parts, however their quality has been proven in a large number of rebuilt units. The repair shop's confidence in this method of repair is supported by a 12-month warranty.

INDEPENDENT HYDRAULIC SHOPS

This example highlights the type of savings that can be made by using a reputable, independently-owned hydraulic shop to carry out your component rebuilds.The motor manufacturer didn't offer this lower-cost solution because the oversize pistons are non-genuine parts and therefore the repair is not to OEM standard. However, just because a repair is not carried out to OEM standard, this doesn't necessarily mean that the rebuilt component won't perform and last like a new one.

Having said that, it is important to distinguish between a properly engineered and proven repair technique that will save you money and a dubious repair that is likely cost you twice as much in the long-term. You can do this by asking the shop carrying out the repair two questions:

- Has the repair technique been proven to be successful in terms of performance and achieved service life?
- Is the repair covered by warranty?

If the repair technique is unproven, or the shop offering the repair is not willing to back it up with a warranty, you would be wise to think twice before proceeding. You need to make an informed decision based on the amount of money the repair technique will save you if it works, versus what it will cost you if it doesn't. The repair shop may be willing to share some of the risk involved in finding out, given that if the technique proves to be successful, they can offer the same solution to other customers. It is a good idea to discuss these issues with the repair shop before making a decision.

AFTERMARKET PARTS

A similar approach should be adopted if a repair shop offers you a lower-cost rebuild, using non-genuine or aftermarket parts. Aftermarket spare parts for hydraulic components sometimes originate from the same factories that supply the component manufacturer and are therefore the same quality. In many cases however, aftermarket parts originate from niche manufacturers and their quality can vary from poor to excellent. For more information on aftermarket parts for hydraulic components, go to www.HydraulicSupermarket.com/aftermarket

If the aftermarket parts are of known quality and the repair shop is prepared to guarantee them, the decision carries minimal risk. If the repair shop hasn't used the aftermarket parts before and therefore their quality is unknown, you need to base your decision on how much money you will save if the parts live up to expectations; what it will cost you if they don't; and whether the repair shop is willing to carry some of the risk involved in finding out.

If you work for a large organization, internal policy may dictate that all repairs to company equipment are carried out to OEM standards, using only genuine OEM parts. I don't have any objection to this type of policy, but because this book is about saving money on the operation and maintenance of your hydraulic equipment, it is necessary for me to point out the alternatives.

DISTRIBUTOR BIAS

While a reputable, independently-owned hydraulic shop can save you a lot of money on hydraulic component rebuilds, there is one trap to watch out for, something I call 'distributor bias'.

Most independently-owned hydraulic shops represent one or more hydraulic component manufacturers as a 'distributor'. This means two things. Firstly, the hydraulic shop has a commercial allegiance to the manufacturers they represent. And secondly, as a distributor, they receive preferential discount from the manufacturer when they purchase replacement components and spare parts for resale to their customers. This is explained in detail in Chapter 17.

It is unusual for a hydraulic shop to be a distributor for more than a few component manufacturers, therefore most hydraulic shops do not receive the same level of discount from every manufacturer. This is the root of distributor bias. It means that if you send a 'Brand-X' pump to a hydraulic shop that is not a 'Brand-X' distributor, it is quite likely that when they dismantle your pump and calculate their cost to rebuild it, they will discover that they can make more money and give you a better deal (on the surface at least) if they sell you a new 'Brand-Y' pump, for which they are a distributor.

The hydraulic shop will tell you something like: "We can repair your old 'Brand-X' pump for $900 or we can sell you a new one made by 'Brand-Y' for $1,100." They know that for the sake of an extra $200, you will buy the new one.

This situation arises as a result of the different levels of discount the hydraulic shop receives from the two manufacturers. In the above example, the hydraulic shop receives distributor discount on the new 'Brand-Y' pump, but because they are not a distributor for 'Brand-X' they receive a lower rate of discount on the spare parts required to rebuild your old 'Brand-X' pump.

Because of these different levels of discount, and therefore profit margins, the hydraulic shop increases their price to repair your 'Brand-X' pump and reduces their price on a new 'Brand-Y' unit. This reduces the difference between the price to repair your old pump and the price of a new one - of a brand that they distribute, to the point where you would be crazy not to buy the new pump.

What's wrong with this you may ask? Well there's not really anything wrong this, other than it may not actually be the best deal for you, the customer, even though it is designed to look that way. The only way to establish if this is a good deal is to get a quote to rebuild your 'Brand-X' pump from a 'Brand-X' distributor.

If a 'Brand-X' distributor were to quote you $650 to repair your old pump, would you still be willing to buy a new 'Brand-Y' unit for $1100? Depending on your price threshold for proceeding with a repair as a percentage of new, which is usually 70%, you probably would not.

CONSEQUENCES OF DISTRIBUTOR BIAS

Distributor bias can cost you money and this is best illustrated with an example. Some time ago, I tried to win the hydraulic repair business of a manufacturing plant situated close to the hydraulic shop where I was working. On one of my visits to this plant, the maintenance manager happened to mention that they had sent one of their pumps to their regular hydraulic shop for repair.

Because I knew that this customer's regular hydraulic shop was not a distributor for this particular brand of pump and my company was, I realized that this was a great opportunity for us to get some business from this customer. So I asked the maintenance manager to give my company a chance to quote on rebuilding this pump. Unfortunately, we didn't get an opportunity to quote.

On a subsequent visit to the plant, I discovered that the original pump had been replaced with a new one of a brand for which the customer's regular hydraulic shop was a distributor. When I inquired why the new pump had been fitted, the maintenance manager advised me that the original pump had been uneconomical to repair. I knew immediately that this maintenance manager had fallen victim to distributor bias.

This was confirmed some time later when, after my company secured this customer's business, we were able to rebuild the remaining, original pumps for around half the price the customer paid for the new pump two years earlier!

This doesn't mean that it is always necessary to send a 'Brand-X' component to a 'Brand-X' distributor to get the best price on a rebuild. Hydraulic shops that specialize in rebuilding hydraulic components often have agreements with a large number of hydraulic component manufacturers, which enable them to buy spare parts at distributor discount, but not complete units. This means that in addition to being price competitive on repairs to components from a large number of manufacturers, a specialist repair shop is much less likely to have distributor bias.

15

Evaluating a hydraulic repair shop

Regardless of whether you decide to use the machine dealer, the hydraulic component manufacturer or an independent for your hydraulic repairs, you should always inspect their facilities first.

Beyond the basics such as a clean, dry environment, you should satisfy yourself that the repairer has the necessary facilities to carry out the majority of the work in-house and test the component once it has been rebuilt.

If the repair shop has to rely on outside suppliers to carry out a large portion of the rebuild process, you are likely to end up paying too much for the repair. This is due to the margins of these suppliers being built into the price you pay for the rebuild. For example, if you send a cylinder to a repair shop that has to use outside suppliers for any welding, machining, rod straightening or chroming, their repair price is likely to be more expensive than a repair shop that has the capability to carry out all these processes in-house.

TEST FACILITIES

It is also essential to ensure that the repairer has the capability to function test the component once it has been rebuilt.This ensures that the component will not only work the way it should when it is fitted to the machine, but will also perform within its design parameters.

This is particularly important with pumps and motors. It is a common misconception that all that is required to return these components to 'as new' condition is to fit a new rotating group. Not only must a new rotating group be fitted, it must be correctly toleranced during assembly so that volumetric efficiency equals or exceeds that of a new unit. The only way to confirm this is to dynamically test the unit under load, on a test bench designed for this purpose.

In the case of variable-displacement pumps and motors, the unit's controls must be adjusted according to the manufacturer's specifications, to ensure optimum performance and therefore machine productivity.These controls are commonly integral to the unit or mounted on it. Many repair shops overlook this step because they lack the necessary test facilities or expertise to carry out these adjustments.

While these adjustments can be carried out in the field after the rebuilt component has been fitted to the machine, assuming the necessary equipment and expertise are available, it is far easier to carry them out in a controlled environment.Therefore, ensure that the repair shop you use has the necessary facilities and expertise to test your component and set its controls before it is dispatched.

16

Common repair rip-offs and how to avoid them

It is difficult to reveal the scams and rip-offs perpetrated by some elements of the fluid power industry without appearing negative. I am not suggesting that the entire industry is crooked, but there are unethical operators out there who will take advantage of you, if you give them an opportunity to do so. The good news is that if you know the pitfalls, you can avoid them. In the case of hydraulic component repairs, common rip-offs include:

REWORKING PARTS AND CHARGING FOR NEW ONES

When a hydraulic component is rebuilt, there are usually some parts that can be successfully re-used after they have been reworked using processes such as machining, honing, lapping and hard-chrome plating. These are money-saving repair techniques and there is nothing wrong with the repair shop employing them, as long as you are not charged the price of a new part!

CHARGING FOR REWORKING PARTS AND NOT REWORKING THEM

An example of this is the re-chroming of cylinder rods. The repair shop advises you that your cylinder rod needs to be re-chromed, when they know very well that the scratches will polish out. You pay for a complete re-chrome when all the repair shop does is polish the existing chrome, so that it looks like new! This is easy money for the repair shop - at the customer's expense.

When a repair shop quotes to re-chrome a cylinder rod, it is always a good idea to ask if the scratches can be removed by polishing. This indicates to the repairer that you are aware that there is sometimes a cheaper solution. In many cases, the rod will actually need to be re-chromed.

As a guide, if the scratches are deep enough to catch your fingernail, they are usually too deep to polish out. Polishing the chrome reduces the finished diameter of the rod. This increases the extrusion gap of the rod seal (the gap between the rod and the land of the seal groove), which reduces the service life of the rod seal.

The only way to confirm that you got the chrome job you paid for, is to check the chrome thickness before and after the cylinder is repaired. To do this you need a coating-thickness gauge. Before a rod is re-chromed, the existing chrome has to be ground off. Each time a cylinder rod is ground, the diameter of the parent metal is reduced and therefore, the thickness of the chrome required to finish the rod to its specified diameter, increases.

This means that if the rod has been re-chromed, the chrome thickness should have increased. If the rod has only been polished, then the chrome thickness (and the rod's finished diameter) will have decreased. For more information on instruments for measuring chrome thickness, go to www.HydraulicSupermarket.com/chrometest

FITTING USED PARTS AND CHARGING FOR NEW ONES

It is sometimes necessary to rebuild a component using one or more used parts. This situation may arise if the component is an old model, which is no longer supported by the manufacturer, or if the required parts are not readily available and you need to get your machine running urgently. Again, there is nothing wrong with this, as long as the used parts are in serviceable condition, you are aware that they are being fitted and you are not charged the price of new parts!

FITTING NON-GENUINE PARTS AND CHARGING FOR GENUINE

As explained in Chapter 14, there is nothing wrong with a repair shop using non-genuine spare parts when rebuilding a hydraulic component, provided that the repairer guarantees the quality of the parts, you are aware they are being fitted and you are not being charged the price of genuine parts!

AVOIDING REPAIR RIP-OFFS

There are three things you can do to ensure that you don't become a victim of one of these costly rip-offs. Firstly, find out exactly what you are paying for. To do this, ask the repair shop for a component inspection report, which details:

- parts that are to be reworked;
- parts that are to be replaced; and
- the quality of each of the parts that are to be replaced (genuine, non-genuine or used).

Any reputable hydraulic repair shop should be happy to provide this information to you. It is also a valuable reference when conducting failure analysis, if the component fails during its warranty period or prior to achieving its expected service life.

Secondly, once you have received this report, arrange to inspect the disassembled component with a representative from the repair shop. Ask the representative to show you the parts that are to be reworked or replaced and ask him to explain why this is necessary. You do have to be guided by the repair shop's expertise, that is after all why your hydraulic component is there in the first place. However, if any of the repair shop's explanations regarding the need to rework or replace certain parts do not make sense, get a second repair quote.

Finally, ask for your old parts back. This is a simple and effective way of ensuring that the parts you have paid for have actually been replaced.

17

How to buy replacement components at the lowest possible price

Rebuilding hydraulic components can significantly reduce the operating cost of your hydraulic equipment. However, there will be occasions when it is not possible or economic to rebuild a component, and it will be necessary to buy a replacement.

When you do have to buy a replacement component, you want to buy it at the lowest possible price. In this Chapter, I explain how the fluid power industry works and show you how you can use this insider information to save money when you purchase new components and parts.

HYDRAULIC SUPPLY CHANNELS AND HOW TO BYPASS THEM

While some hydraulic component manufacturers sell their products directly to end-users, most only sell through manufacturers of hydraulic machinery (OEM's) or fluid power distributors. This means that when you need a replacement hydraulic component, you have to buy it from either the machine dealer or a fluid power distributor.

You will usually get a better price on a replacement hydraulic component if you are able to buy it from a fluid power distributor. However, this is not always as easy as it might seem. This is because OEM's usually do all they can to control the distribution and sale of spare parts and components for the machines they build. This is particularly true in the case of mass-produced hydraulic machines. The OEM knows that if you can identify the make and model of the hydraulic pump fitted to your skid-steer loader for example, you will be able to shop around for the best price on a replacement pump and as a result there is a good chance they will lose the business.

In an effort to prevent this from happening, OEM's identify the hydraulic components fitted to their machines with their own part numbers, which in most cases are meaningless to a fluid power distributor. Therefore the first thing you need to do, in order to get the best possible deal on a replacement component, is to identify its manufacturer and model code. Once you have this information, not only can you buy the component from a fluid power distributor who will often sell it to you cheaper than the machine dealer, you are also in a good position to negotiate the price down. To do this, you need to know how the industry works and understand its pricing structure.

INDUSTRY PRICING STRUCTURE AND DISTRIBUTOR MARGINS

Hydraulic component manufacturers set recommended retail prices for the products they manufacture and then sell these products to their distributor network at a discount to the retail price. While the rate of this discount varies, there are two main categories: distributor and reseller. Distributor discount can range between 35% and 55% off retail price and reseller 20% to 30% off retail price.

This means that if a pump has a retail price of $1,000 and distributor discount from the manufacturer is 50%, a distributor's cost price is $500. Therefore, if a distributor sells the pump to a customer for full retail price, their profit is $500: a margin of 100% on their cost price! This may seem like an obscene amount of profit for the distributor. In theory at least, this margin is given to the distributor to cover costs associated with carrying inventory and promoting and selling the manufacturer's products.

Note that the cost of the same pump to a reseller, who receives 25% discount off retail price, is $750. Therefore, if a reseller sells the pump to a customer for full retail price, their profit is $250: a margin of 33% on their cost price. As you can see, a reseller makes a much smaller profit margin than a distributor. This is important to know, if you want to buy a replacement component at the lowest possible price.

Let's say you want to negotiate a 10% discount off the retail price of this pump. This means that the price you will pay is $900. If a distributor sells the pump to you for $900, their profit is $400: a margin of 80% on their cost price. If a reseller sells the pump to you for $900, their profit is $150: a margin of 20% on their cost price. As you can see, you stand a better chance of getting a 10% discount from a distributor, because they have a bigger profit margin to play with. A distributor can give you 10% discount and still make a healthy profit.

DISTINGUISHING DISTRIBUTORS FROM RESELLERS

It is not always easy to distinguish a distributor from a reseller. Just because a company includes a manufacturer in their product line, it doesn't necessarily mean they have distributor status with that manufacturer. They may be a distributor for one or two of the manufacturers in their product line and a reseller for the others. As a rule, if a fluid power company has distributor status with a manufacturer, the words 'Authorized Distributor' will be used in association with that manufacturer's name.

UNDERSTANDING YOUR BUYING SITUATION

Fluid power distributors are in business to make money and so they don't give away their profit margin easily. Getting a distributor to give you a discount is easier if you understand how they think. But first, you need to understand your buying situation.

From a distributor's point of view, you, the customer, will fall into one of three buying situations when you purchase hydraulic components. These are: straight re-buy; modified re-buy; and new buying situation. To explain these buying situations and what they mean to a fluid power distributor, I will use a hydraulic log splitter as an example.

Let's say you are going to build a hydraulic log splitter. You have an engine, you plan to fabricate the frame yourself, but you need to buy all the hydraulic components. You know you will need a pump, control valve, cylinder and a few other items, but you are not sure how to size these components according to the speed and force required. So you call a fluid power distributor, explain what you want to do and ask them to quote to supply what you need.

Before the distributor can quote, an application engineer has to design the hydraulic circuit and specify the components required. This may take several hours. The distributor doesn't charge you for the application engineer's time, but they will try to get full retail price for the components you need for your log splitter to cover this cost. This is a new buying situation. New buying situations usually require a lot of technical work on the part of the distributor and so when you are in this buying situation, the distributor will be most reluctant to give you a discount.

Take another scenario. Let's say you already have a log splitter, but the machine is quite old and the pump is worn out. When you go to buy a replacement pump, you learn that the manufacturer of the pump has gone out of business. You therefore need an equivalent pump that matches the obsolete unit as closely as possible, so you take the old pump to your nearest fluid power distributor.

Before the distributor can quote you a price on an equivalent pump, a sales engineer has to identify all the specifications of the existing unit, such as shaft, mounting, ports and displacement, and then cross-reference this information to find a suitable alternative. This may take an hour or more of the sales engineer's time. Again, the distributor doesn't charge you for the sales engineer's time, but they will try to get full retail price for the alternative pump, to cover this cost. This is a modified re-buy situation. Modified re-buy situations require some technical work on the part of the distributor and therefore when you are in this buying situation the distributor will be reluctant to give you a discount.

In a third scenario, let's say the pump on your existing log splitter is worn out, so you take down the details from the pump's identification tag, such as the manufacturer, model code and serial number. You call your nearest distributor for this manufacturer and ask for a price on a 'Brand-X' pump, model code PF1-234/XYZ. The sales person enters this model code into their computer and is almost instantly able to quote you a price on a replacement pump. This is a straight re-buy situation. Straight re-buy situations require no technical work on the part of the distributor. This type of transaction is easy money for the distributor and therefore when you are in this buying situation, you are in the best position to get a discount.

CREATING COMPETITIVE TENSION

One thing to remember, regardless of which buying situation you are in, is that recommended retail prices are exactly that. There is nothing to prevent a distributor from selling a component to you below the manufacturer's recommended retail price. If there was, it could be construed as price-fixing, which is illegal in many countries.

This means that if, in the last example, you were to get prices on the 'Brand-X' pump from three, different distributors and all three distributors quoted you the recommended retail price, there is nothing stopping you from creating competitive tension to get a better deal. You could call one of the distributors and say something like: "I have three quotes on this pump and they are all within a few dollars of each other. If you will sell it to me for $XXX (your target price) I'll give you the business." If the first distributor doesn't respond favorably, you can always call the other two and put the same proposition to them. There is a good chance that one of them will play ball and give you a discount, particularly as this is a straight re-buy situation.

BUYING REPLACEMENT HYDRAULIC COMPONENTS

If you need a replacement component for an existing hydraulic machine, you will be in either a straight re-buy or modified re-buy situation. If the component has its manufacturer's identification tag attached, complete with model code, this is a gift. You are automatically in a straight re-buy situation.

If the component can only be identified by the machine manufacturer's part number, and you want to buy a replacement from a fluid power distributor, then you are effectively in a modified re-buy situation. This is because the distributor has to identify the component's manufacturer and specifications and convert this information into a model code before they can quote you a price on a replacement. Even though the distributor will probably quote you full retail price on a replacement, this price is likely to be cheaper than buying the same component from the machine dealer.

HOW TO GET TO A STRAIGHT RE-BUY SITUATION

As you now know, negotiating a discount from a distributor is easier if you are in a straight re-buy situation. The catch is, the distributor who has identified your old component is unlikely to give you the model code, because they know this makes it easy for you to shop around for the best price. But there are other ways to get the model code you need.

If you know another operator or company that owns the same machine you need the replacement component for, you could ask them if they have replaced this component with one that wasn't purchased from the machine dealer. If so, they will probably be happy to give you the model code from the component's identification tag.

If you have a reasonably good idea of what you are looking for, you can identify the component yourself, using information that is freely available on the Internet. This involves first measuring and identifying the component's physical attributes such as shaft type, mounting flange, ports and displacement. You then need to match these variables to the dimensional and technical data contained in the manufacturer's product catalog. The product catalog will show you how to compile a model code (sometimes called an order code) that corresponds to the component you want to replace. Once you have this information, you are in a straight re-buy situation.

Sounds easy, but if you don't know who the manufacturer is or what the component does, this can be a difficult task. In this case, it is easier to let a fluid power distributor do the work and earn their profit margin in the process. But if you do have the time and inclination, a comprehensive list of hydraulic component manufacturers' product catalogs is available at www.HydraulicSupermarket.com/products and information to help you identify standard shaft sizes, mounting flanges, port sizes, thread types and more, is available at www.HydraulicSupermarket.com/technical

If neither of the above options yield a model code, it would be wise to get another quote anyway. If the manufacturer of the component is not obvious, the distributor that identified it will probably tell you this much with a little prompting. Once you know who the manufacturer is, you can take the component to one or two distributors to get alternative prices. Unfortunately, this won't give you the same leverage as a straight re-buy situation.

The good news is, once you have bought the component through a fluid power distributor, all the details you need for a straight re-buy situation next time you replace the component, will be on the component's identification tag.

OEM SPECIALS

Having said all that, it is not always possible to buy the replacement component you need from a fluid power distributor. OEM's sometimes fit hydraulic components to their machines that are manufactured with a unique difference, known as 'OEM specials', so that even if you do identify the make and model of the component, the only way you can buy an identical unit is through the machine dealer.

The difference may be something subtle such as the shaft type or the orientation of the ports, however this is usually enough to make it either impossible or uneconomic to fit a similar component from the manufacturer's standard product line. Some large OEM's also manufacture their own hydraulic components. As with OEM specials, these components are a captive market for the machine dealer.

SHOPPING BETWEEN COMPETING BRANDS

Another thing you should know about fluid power distributors is that they always relish an opportunity to sell a component that will replace an existing component manufactured by a competitor. In many cases, you can use this fact to get a better deal on a replacement component.

Consider the example we discussed earlier. You need a 'Brand-X' pump, and you have the model code. You have three quotes from 'Brand-X' distributors, but these prices look expensive. So you call a 'Brand-Y' distributor and ask if they can offer you a 'Brand-Y' equivalent.

If the 'Brand-Y' distributor can supply a suitable equivalent, they are likely to offer it to you at a discounted price. This is because you are buying the 'Brand-Y' pump as a replacement for a competing brand.

Assuming the price for the 'Brand-Y' pump is cheaper than the price you were quoted for the 'Brand-X' unit and the two brands are similar in quality, you can use this as leverage to get a better deal from the 'Brand-X' distributor.

You can do this by calling one of the 'Brand-X' distributors and saying something like: "Thanks for your quote, but I can buy an equivalent 'Brand-Y' pump cheaper, and unless you improve your price I am going to throw my 'Brand-X' pump in the scrap bin and fit a 'Brand-Y' unit in its place." The 'Brand-X' distributor will not want to lose the sale to a competing brand and therefore is likely to reduce their price. But keep in mind that if the two brands are not similar in quality, your leverage will be diminished because the 'Brand-X' distributor will advise you that the 'Brand-Y' pump is cheaper because it is inferior in quality.

INTERCHANGEABILITY BETWEEN BRANDS

Some hydraulic components are manufactured to industry standards. Examples include gear pumps manufactured to DIN or SAE standards, cylinders manufactured to NFPA standards, and subplate-mounted valves manufactured to CETOP standards. These standards ensure interchangeability between manufacturers.

If the component you are replacing is not manufactured to a recognized standard, locating an identical component from an alternative manufacturer may prove difficult. In which case, the extent and cost of any modifications required will determine if fitting a component from an alternative manufacturer is an economic proposition.

If in the above example, you decided to go with the 'Brand-Y' pump, you would first need to confirm that it was directly interchangeable with the 'Brand-X' pump, or if it wasn't, what the differences were and how much any necessary modifications would cost. If the 'Brand-Y' pump was $200 cheaper but you had to spend $150 to modify the intake line, then it is probably not worth the trouble.

SAVING MONEY ON SURPLUS COMPONENTS

A frequently overlooked option that can yield significant savings on the replacement cost of a hydraulic component is to buy a surplus unit. This is particularly the case if the component you need is no longer in production or is very expensive.

It is not uncommon for fluid power distributors and owners of hydraulic machinery to accumulate inventory of hydraulic components, which have become obsolete or surplus to their requirements. The owners of this stock are often willing to sell these components below replacement cost.

The challenge with this option is finding the component you need. When you consider all of the different types, sizes and variations of hydraulic components and the number of manufacturers producing them, finding the exact component you require can be like looking for a needle in a haystack.

The goods news is that you can dramatically increase your chances of finding a surplus component that matches your requirement by using the Internet. HydraulicSupermarket.com operates an online marketplace, which is dedicated to the trading of surplus hydraulic components. This marketplace allows sellers of surplus components to submit details of items available for sale to an online database. Buyers looking for a surplus component can search this database and when a match is found, submit an offer to the seller. To search for a surplus hydraulic component using this unique facility, go to www.HydraulicSupermarket.com/market

18

How to safeguard new component warranty and get free repairs after warranty has expired

Whenever you replace a hydraulic component with a new or rebuilt unit, the supplier of the component will warrant it to be free from defects in materials and workmanship for a set period of time (the one possible exception to this rule is if you purchase a surplus component). The length of this warranty period can vary, but 90 days from date of purchase is usually the minimum.

It is important to understand that this warranty is not unconditional. If you have ever taken the time to read the warranty conditions for a hydraulic component, you would be aware that there is a list of circumstances that, if deemed to have caused the component to fail, can result in rejection of a warranty claim. These include:

- Improper storage or handling.
- Interference or tampering with the component while in storage, such as the removal of parts, even if these parts are replaced prior to the unit being put into service.

- Failure to follow proper commissioning or start-up procedures.
- Incorrect setting or adjustment.
- Contaminated hydraulic fluid.
- Incorrect fluid viscosity.
- Damage caused by incorrect adjustment of the machine's hydraulic or electronic system.
- Damage caused by faulty components in the machine's hydraulic or electronic system.
- Improper machine operation.

Warranty claims usually result in win/lose outcomes. If a component fails during its warranty period, either the supplier or the customer has to pay to fix it. Who pays depends on the cause of the failure and who is responsible for it.

SAFEGUARDING NEW COMPONENT WARRANTY

If you implement a preventative maintenance program, such as the one outlined in Part I of this book, and a component happens to fail during its warranty period, then it is unlikely that the warranty claim will be rejected. But there are specific things you can do to ensure that you are covered if a warranty situation eventuates.

One of the most common causes of premature failure of hydraulic components is incorrect commissioning when the component is installed. A large number of warranty claims are rejected on this basis. The best way to eliminate this responsibility is to hire the component supplier's service technician to fit and commission the component, and also check and adjust the hydraulic system's settings.

While this will involve some additional cost, it can be well worth the expense. If a warranty claim does eventuate and the failure is attributed to incorrect commissioning or incorrect adjustment of the hydraulic system's settings, responsibility for the failure falls back on the supplier of the component.

If it is not practical to get the supplier to fit and commission the component, the next best thing is to request that they provide a detailed commissioning procedure. This procedure should be carefully followed and properly documented, so that it can be referred to in the event of a warranty claim. If a warranty claim does eventuate and the failure is attributed to incorrect commissioning, then as long as you can demonstrate that this commissioning procedure was followed, at least some of the responsibility for the failure falls back on the supplier.

Contamination is another common reason for rejecting warranty claims. It is therefore wise to take a fluid sample prior to fitting replacement components. This ensures that you will have a reference cleanliness level for the fluid in the system, should contamination be used as a basis for rejecting a subsequent warranty claim.

GETTING FREE REPAIRS AFTER WARRANTY HAS EXPIRED

In some cases, when a hydraulic component fails prematurely it is possible to get free repairs even after the warranty period has expired. If the component itself or the hydraulic circuit in which it operates has a design fault which influenced the failure, then it may be possible to get the component repaired at no cost under a policy or goodwill concession from the component or machine manufacturer.

These concessions are an extension of warranty on a case-by-case basis.This is best illustrated with an example. Some time ago I was involved in failure analysis of a variable-displacement pump that had failed prematurely, after its warranty period had expired.The cause of the failure was over-pressurization.

The pump's primary over-pressure protection device was the pressure limiter on its displacement control and its secondary over-pressure protection device was a relief valve. Analysis revealed that the pump's pressure limiter had ceased to function as a result of contamination, which was isolated to this device rather than systemic. This meant that the pump had to rely on its relief valve to limit system pressure. But this relief valve was too small to handle the rated flow from the pump. As a consequence, there was a large pressure rise across the relief valve, which resulted in the over-pressurization and subsequent failure of the pump.

The customer complained to the machine manufacturer on the basis that the pump would not have failed if the relief valve fitted to the system had been capable of handling the rated flow of the pump. To cut a long story short, the machine manufacturer eventually agreed to pay the cost of repairing this pump. This was a policy concession from the manufacturer, given without any admission of fault or liability on their part.

The important thing to understand about policy or goodwill concessions is that you generally don't get them unless you ask for them. Other owners of the same machine, who didn't complain, could have paid for the same repairs that this particular owner got for free.

This example also reinforces the importance of conducting thorough failure analysis whenever a hydraulic component fails prematurely.

PART III CONCLUSION

The price you pay for replacement components, whether new or rebuilt, has a significant impact on the operating costs of your hydraulic equipment. By applying the principles outlined in the preceding Chapters, you will avoid the rip-offs and purchase new or rebuilt components at the most economical price.

Glossary

ABSOLUTE FILTER RATING - the diameter in microns of the largest, spherical particle that will pass through a filter element under controlled conditions.

ABSOLUTE PRESSURE - a pressure scale with a zero point at perfect vacuum (total absence of atmospheric pressure).

ABSOLUTE VISCOSITY - the ratio of a fluid's shearing stress to shear rate.

ACCUMULATOR - a device in which hydraulic fluid is stored under pressure.

ACCUMULATOR EFFECT - the increase in volume of a hose or pipe as pressure increases.

ACTUATOR - a device that converts hydraulic energy into mechanical energy e.g. motor or cylinder.

AERATION - air contamination of the hydraulic fluid.

AFTERMARKET PARTS - parts manufactured or marketed as substitutes for those supplied by the Original Equipment Manufacturer.

ANTI-CAVITATION VALVE - a device that enables the make-up of additional fluid, once the fluid demanded in part of a circuit has started to exceed the available supply, thereby preventing cavitation.

ASTM - American Society for Testing and Materials.

ATMOSPHERIC PRESSURE - pressure exerted by the atmosphere, 14.7 PSI at sea level.

BETA RATIO - Filter efficiency ratio defined according to ISO 4572 and expressed as the Beta ratio or rating (β) for a given particle size (χ). The Beta ratio value is derived as follows:

$$\beta\chi = \frac{\text{number of particles of size } \chi \text{ upstream of the filter}}{\text{number of particles of size } \chi \text{ downstream of the filter}}$$

BREATHER - a device that permits air to move into and out of the reservoir as fluid volume changes, thereby maintaining the reservoir at atmospheric pressure.

BSI - British Standards Institution. BSI is the British institute for standardization.

CAVITATION - a term used to describe the formation, growth and implosion of vapor or gas-filled cavities in a liquid. These cavities are formed when the static pressure of the liquid falls below its vapor pressure at ambient temperature. When a cavity in the liquid is subjected to rapid pressure rise, it implodes violently. This can cause metal erosion in hydraulic components.

CENTISTOKE - a unit of kinematic viscosity.

CETOP - Comité Européen des Transmissions Oléohydrauliques et Pneumatiques. CETOP is the European fluid power association.

CHARGE PRESSURE - the pressure to which the loop is supercharged in a hydrostatic transmission.

CHROMING - to electroplate with hard-chrome.

CIRCUIT - an arrangement of interconnected components.

CLOSED CENTER - a circuit in which pump output is prevented from circulating (closed) to the reservoir when the directional control valve is in the center or neutral position.

CLOSED CIRCUIT - a circuit in which fluid from the motor outlet flows directly to the pump inlet, without returning to the reservoir. Sometimes called closed loop.

COMPONENT - see hydraulic component.

CONDUCTOR - a carrier of fluid, i.e. tube, pipe or hose.

CONTAMINATION - any matter that impairs the function of the hydraulic fluid, especially insoluble particles, air and water.

COUNTERBALANCE VALVE - a valve that maintains resistance to flow in one direction and permits free flow in the other. Used in applications where actuators support suspended loads. Prevents uncontrolled movement of the load.

CYLINDER - a device that converts fluid power into linear force and motion.

DIESEL EFFECT or DIESELING - the explosion of a mixture of air and hydraulic fluid when pressurized. Occurs in a hydraulic cylinder when air is drawn past the rod seals or because of failure to bleed air from the cylinder during commissioning.

DIN - Deutsches Institut für Normung. DIN is the German institute for standardization.

DIRECTIONAL CONTROL VALVE - a valve that controls the direction of fluid flow in a hydraulic circuit.

DISPLACEMENT - the quantity of fluid that passes through a pump, motor or cylinder in a single revolution or stroke.

DOUBLE-ACTING CYLINDER - a cylinder in which fluid pressure can be applied in either direction.

DRAIN LINE - a tube, pipe or hose conducting internal leakage from a component to the reservoir.

EFFICIENCY - the ratio of output to input, usually expressed as a percentage.

FILTER - a device that removes insoluble contaminants from hydraulic fluid.

FILTER BYPASS VALVE - a valve that allows fluid to bypass a filter element when the element clogs. Prevents the element from being subjected to excessive pressure drop.

FLOAT VALVE - a valve that allows an attachment such as the boom or arm on a hydraulic excavator to be lowered under its own weight.

FLOODED HOUSING - pump case and inlet are connected internally and are below reservoir fluid level.

FLOODED INLET - pump inlet is below reservoir fluid level.

FLOW CONTROL VALVE - a device that controls the rate of fluid flow.

FLOW RATE - the volume of fluid moving through a conductor per unit of time.

FLOW-TESTER - an instrument that comprises a flow turbine for measuring flow rate, an adjustable orifice that is used to increase the resistance to flow (load valve) and a pressure gauge, which measures pressure upstream of the load valve. Also called a flow-meter.

FLUID - a liquid or gas (see also hydraulic fluid).

FLUID CLEANLINESS LEVEL - the quantity and size of insoluble particles present in the hydraulic fluid. Usually defined according to ISO 4406.

FLUID POWER - the transmission of power using a liquid or a gas i.e. hydraulics and pneumatics.

FLUSHING - 1) the process of circulating hydraulic fluid through an external filter to remove particle contamination. 2) positive extraction and therefore circulation of fluid through a component or circuit to prevent localized heating of the fluid.

HEAT EXCHANGER - a device that transfers heat from one fluid (liquid or gas) to another.

HYDRAULIC COMPONENT - a hydraulic assembly such as a pump, motor, cylinder or valve.

HYDRAULIC FLUID - a liquid specially formulated for the transmission of power in a hydraulic system.

HYDROSTATIC TRANSMISSION - a variable-displacement hydraulic pump and a fixed or variable-displacement hydraulic motor, operating together in a closed circuit. Also called a hydrostatic drive.

INTAKE LINE - a tube, pipe or hose conducting fluid from the reservoir to the pump inlet.

ISO - International Organization for Standardization.

KINEMATIC VISCOSITY - a fluid's absolute viscosity divided by its density.

LAND - the ridge of metal either side of a seal groove.

LAPPING - the use of abrasive slurry to produce a flat, smooth finish and therefore even contact between metal parts.

LINE - a tube, pipe or hose that conducts hydraulic fluid.

LOAD CONTROL VALVE - a device that prevents uncontrolled actuator movement due to gravity e.g. counterbalance valve.

MF - abbreviation for motor, fixed-displacement.

MICRON - unit of measurement. One millionth of a meter.

MILLILITER - unit of volume. One thousandth of a liter.

MILLIMETER - unit of measurement. One thousandth of a meter.

MOTION CONTROL VALVE - a device that prevents uncontrolled actuator movement due to inertia.

MOTOR - a device that converts fluid power into rotary force (torque) and motion.

MV - abbreviation for motor, variable-displacement.

NAS - National Aerospace Standard.

NFPA - National Fluid Power Association.

NOMINAL FILTER RATING - an arbitrary particle blocking rating, nominated by the filter manufacturer.

OEM - Original Equipment Manufacturer. A manufacturer of machinery or components.

OPEN CENTER - a circuit in which pump output is allowed to circulate (open) to the reservoir when the directional control valve is in the center or neutral position.

OPEN CIRCUIT - a circuit in which fluid from the actuator outlet returns to the reservoir before entering the pump inlet.

ORIFICE - a restriction to fluid flow, the length of which is small, relative to its diameter.

PEAK-PRESSURE METER - an electronic instrument that records the maximum, instantaneous pressure in a hydraulic circuit.

PF - abbreviation for pump, fixed-displacement.

PILOT LINE - a tube, pipe or hose conducting fluid at pilot pressure.

PILOT PRESSURE - auxiliary pressure used to actuate or control hydraulic components.

PPM - parts per million.

PRESSURE COMPENSATOR - a displacement control for variable pumps in which displacement is reduced once a pre-set pressure is reached.

PRESSURE DROP - the difference in pressure between any two points in a circuit. There is always pressure drop when there is flow, due to friction between the fluid and conductor.

PRIME MOVER - the power source driving a pump, i.e. electric motor or internal combustion engine.

PV - abbreviation for pump, variable-displacement.

RAM - a single-acting cylinder in which fluid pressure acts on the cross-sectional area of the rod (no piston).

RELIEF VALVE - a device that diverts pump flow to the reservoir once a pre-set pressure is reached. Limits system pressure to a predetermined, maximum value.

RESERVOIR - a vessel that stores fluid in a hydraulic system. Also called a tank.

RETURN LINE - a tube, pipe or hose conducting fluid from the actuator outlet to the reservoir.

ROTATING GROUP - an assembly comprising the rotating parts of a pump or motor.

SAE - Society of Automotive Engineers.

SERVICEABLE - in functional condition, able to render service.

SERVO VALVE - a device that controls the direction and flow rate of fluid in proportion to an electrical input signal, and incorporates an internal feedback mechanism for high accuracy.

SILT - insoluble particles less than 5 microns in diameter.

SINGLE-ACTING CYLINDER - a cylinder in which fluid pressure can be applied in one direction only.

SPOOL - a cylindrically-shaped part, which moves to direct flow through a component.

STRAINER - a coarse filter commonly used as a pump inlet filter.

SUB-PLATE - an auxiliary mounting that provides a means of connecting lines to a component.

SUCTION LINE - a tube, pipe or hose conducting fluid, at below atmospheric pressure, from the reservoir to the pump inlet.

TANK - a vessel that stores fluid in a hydraulic system. Also called a reservoir.

VACUUM - the partial or complete absence of atmospheric pressure.

VALVE - a device that controls fluid direction, pressure or flow rate.

VAPOR PRESSURE - the pressure at which the liquid and gaseous states of a fluid are in equilibrium at a given temperature.

VG - viscosity grade (of hydraulic fluid).

VISCOSITY - the internal friction of a fluid i.e. its resistance to flow.

VOLUMETRIC EFFICIENCY - the actual output of a pump in GPM, divided by its theoretical output and expressed as a percentage.

Index